#CommActUpdate

A Communications Law Fit for the
Digital Age

#CommActUpdate

A Communications Law Fit for the Digital Age

Randolph J. May
Seth L. Cooper

The Free State Foundation

#COMMACTUPDATE by Randolph J. May and Seth L. Cooper

Published by The Free State Foundation, Inc., Potomac, MD

Copyright © 2017 by Randolph J. May and Seth L. Cooper

Author services provided by Pedernales Publishing, LLC
www.pedernalespublishing.com

Library of Congress Control Number: 2017913663

ISBN-13:   978-0-9993608-2-8   Paperback edition
            978-0-9993608-0-4   Digital edition

Printed in the United States of America

For my wife Laurie, whose support is always unwavering, and my children, Joshua and Brooke, from whom I draw never-ending inspiration

For my wife Gretchen, with whom I share love and life, and my children, Elysia, Lorelei, Calista, and Soren, in whom I delight

CONTENTS

PREFACE

After the 2016 elections, there is a real opportunity to achieve a much-needed, meaningful modernization of our nation's communications laws. President Donald Trump made elimination of unnecessary regulations a top priority in his campaign. And the 2016 congressional elections retained Republican majorities in the Senate and House of Representatives. Many of those who were elected ran on platforms that put reducing the costs and burdens of unnecessary regulations at or near the top of their campaign platforms. Moreover, both chambers of Congress have leaders that have supported previous legislative attempts to get rid of obsolete regulations and to modernize the Communications Act.

Most of the laws now governing communications services, now including the Internet, are twenty or even eighty years old. The premises underlying these laws are far removed from the breakthroughs in digital, broadband, and wireless technologies we enjoy today. Reform cannot wait much longer – and should not have to wait much longer. Opportunities for eliminating or curtailing outdated regulatory restrictions embodied in law are relatively rare. This is true even when widely available evidence demonstrates those restrictions no longer comport with marketplace realities or with the rationales originally proffered to justify their adoption.

The new President and new Congress are now in a position to replace the outdated Communications Act of 1934 – which was shaped with a heavy bias toward government regulation of perceived monopolies for analog services – with a modernized Communications Act that is based on a preference, whenever possible, for marketplace competition rather than government regulation. A clean-slate reform-minded overhaul of the Communications Act that regards market competition as the default norm – while also maintaining circumscribed regulatory authority to address specific anticompetitive harms to consumers caused by market failures, along with targeted measures to help ensure universal service – should have bipartisan appeal. The Trump Administration and the 115th Congress should take the initiative and proactively pursue adoption of a modernized Communications Act.

The Federal Communications Commission's (FCC) highly controversial 2015 "net neutrality" order imposing public utility-style regulation on broadband Internet access services also points to the pressing need for a modernized Communications Act. Broadband infrastructure investment has declined in the wake of the FCC's order, which also had the effect of stripping the Federal Trade Commission (FTC) of its authority over broadband Internet service providers' online privacy practices. Under the leadership of new Chairman Ajit Pai, the FCC initiated a new rulemaking proceeding in May 2017 to restore a light-touch policy toward broadband services. If ultimately adopted by the FCC, the proposed rulemaking would return jurisdiction over broadband Internet providers' consumer privacy practices to the FTC. Yet, a return to a lighter touch regulatory regime than that which now prevails would still leave broadband services vulnerable to renewed regulatory overreach by the FCC in the future.

Even if Congress cannot pursue a more comprehensive overhaul of the Communications Act in the near-term, it should act promptly to provide a long-lasting resolution to these net neutrality and online privacy policy issues. It should adopt legislation to restore and sustain Internet freedom, innovation, and investment, while also providing a narrowly circumscribed means for the FCC to address specific instances of demonstrated consumer harm and anticompetitive conduct by broadband service providers. Similarly, Congress should adopt legislation to re-establish the FTC as the agency responsible for consumer privacy across all online platforms and services.

While it certainly may make sense for Congress to act more quickly in a targeted fashion to resolve the net neutrality disputes that are now more than a decade old, a more comprehensive clean-slate reform of the nation's communications laws offers a superior approach for Congress to resolve broadband and online privacy policy for the long-term. Indeed, a modernized Communications Act is essential to providing a free market, rule of law-oriented framework governing communications services for the Digital Age.

When meaningful reform of our nation's communications laws finally has been achieved, it won't matter much whether this effort initially was denominated an "update," a "rewrite," or something else. What matters is that, at the end of the day, our nation's communications laws are fit for the Digital Age, not mired in obsolete Analog Age suppositions.

In view of the present opportune moment, the Free State Foundation has prepared *#CommActUpdate: A Communications Law Fit for the Digital Age*. Preparation of this book also closely follows on the heels of the Free State Foundation's Tenth Anniversary. The Free State Foundation (FSF) is a non-profit, nonpartisan think tank dedicated to promoting understanding of free market, limited government, and rule of law principles, especially with regard to communications laws and policies. Over the past decade, Free State Foundation scholars have been a leading voice for communications policy reform, consistently advocating the replacement of the existing legacy regulatory regime that governs communications services with a competition-focused framework that matches today's dynamic market realities.

During 2014-2015, Free State Foundation scholars, including

a significant number of the members of the Foundation's Board of Academic Advisors, collaborated in responding to the series of White Papers issued by the House Committee on Energy and Commerce in connection with a project to modernize the Communications Act. FSF's six responses to the House Committee's White Papers comprise a large portion of this book, which sets forth fundamental principles and specific policy prescriptions that should form the basis for an overhaul of federal communications laws – a "#CommActUpdate" as the House Committee put it.

Please note, as does each individual response, that not all of the contributors to FSF's responses agreed on all points. Nevertheless, all agreed on the need for meaningful reform of the Communications Act in the direction of less reliance on regulation and more reliance on marketplace competition as a means of protecting consumers from harm and enhancing their welfare.

We hope this monograph will provide a helpful guide for the task of communications reform that is now set before the new President and new Congress.

· · ·

Finally, we extend our appreciation to Kathee Baker, the Free State Foundation's Communications and Events Coordinator, for all her editorial and other assistance in the preparation of this manuscript. Kathee exhibited her usual dedication, helpful comments, and good cheer throughout the process. Senior Fellow Ted Bolema, who joined the FSF team only recently, also contributed helpful editorial suggestions as we were completing the manuscript.

Randolph J. May
Seth L. Cooper
Rockville, Maryland
August 2017

INTRODUCTION: THE NEED FOR A MODERNIZED COMMUNICATIONS ACT AND THE WAY FORWARD

America needs a new communications law and new communications policies fit for the Digital Age. Advances in digital technology are rapidly reshaping the market and the static premises of industrial age policy are fast fading into obsolescence. Yet too much of the industrial age regulatory framework remains entrenched. The dynamic communications technologies of the early 21st Century are being curbed by government regulatory controls rooted in the late 19th Century and early 20th Century. Market innovation is now constrained by regulatory restrictions that have long outlived their original rationales.

As innovative products and services are inhibited by rules intended for technologies of an earlier era, the need for a modernized law that comports with the new era of digital communications grows increasingly urgent. That urgency is all the more pronounced given the Federal Communications Commission's (FCC) historical pro-regulatory bias. The FCC, especially over the last few years, has proven itself very reluctant or unwilling to relax or remove regulatory restrictions that are now obsolete. Instead, the FCC has proactively sought to regulate new digital communications services in competitive markets without sufficient analytical or statutory justification. Thus, America runs an increasing risk of chilling innovation and investment, impeding market competition, and harming consumer welfare by delaying adoption of a modernized Communications Act.

A vibrant future for digital age communications services ultimately requires comprehensive reform that is pro-innovation, pro-free market, and pro-consumer. But if, for whatever reason, Congress cannot pursue a more comprehensive overhaul of the Communications Act in the near-term, it should act promptly to pursue a satisfactory legislative resolution to the long-running disputes and litigation – running over a decade now – regarding so-called net neutrality regulations. In this regard, such legislation should preserve Internet freedom, innovation, and investment, while also providing a narrowly circumscribed means for the FCC to address specific instances of demonstrated consumer harm and anticompetitive conduct by broadband Internet service providers. By the same token, Congress should adopt legislation to reestablish the FTC as the agency responsible for protecting consumer privacy across all online platforms and services.

Legacy Regulatory Assumptions Increasingly Are Irrelevant to Today's Competitive Communications Marketplace

The communications marketplace of the Digital Age is characterized by an abundance of new products and services that, if not restricted by regulators, offer tremendous sources of value to consumers. Innovative digital technologies hold the promise of eviscerating nearly all traditional telephone/cable/satellite/broadcast service distinctions. Providers operating in once-disparate service sectors increasingly can compete with each other in a convergent, Internet Protocol (IP)-based communications services market – again, if the regulators allow. Consumers are increasingly able to access and enjoy the same products and services through different digital communications platforms. Indeed, the potential for increased competition and consumer choice brought about by gigabit broadband networks,[1] next-generation 5G wireless spectrum technologies,[2] multi-faceted Internet-connected smart devices,[3] emerging digital media services,[4] and cutting-edge software applications are only beginning to be realized.[5]

Despite these breakthroughs and their benefits to consumers, in early 2017 the digital communications market remains largely in the grip of the strictures of the Communications Act of 1934, with antecedents reaching back to the Interstate Commerce Act of 1887. While the Telecommunications Act of 1996 did update the 1934 Act in some significant ways, it did not sufficiently dictate an end to the traditional command-and-control regulatory structure. So, the laws that govern digital communications technologies still owe much more to the era of copper-wire telephones, telegraphs, and railroads than they owe to the Digital Age's Internet Protocol (IP) technology, the World Wide Web, and high-speed broadband services. The existing framework continues to invite intrusive market oversight and control by government regulators.

Digital communications have been subjected to a regulatory

---

1   CableLabs, "Cable Broadband Technology Gigabit Evolution" (Fall, 2016), available at: http://www.cablelabs.com/wp-content/uploads/2017/03/Cable-Broadband-Technology-Gigabit-Evolution.pdf; Andrew Burger, Telecompetitor, "Report: North America Dominates Global Gigabit Deployments" (August 15, 2016), available at: http://www.telecompetitor.com/report-north-america-dominates-global-gigabit-deployments/.

2   *See, e.g.*, Accenture Strategy, "Smart Cities: How 5G Can Help Municipalities Become Vibrant Smart Cities" (January, 2017), available at: https://newsroom.accenture.com/content/1101/files/Accenture_5G-Municipalities-Become-Smart-Cities.pdf.

3   *See, e.g.*, Verizon, "State of the Market: Internet of Things 2016" (April 2016), available at: https://www.verizon.com/about/sites/default/files/state-of-the-internet-of-things-market-report-2016.pdf.

4   *See, e.g.*, Leichtman Research Group, Press Release: "82% of U.S. TV Households Have a DVR, Netflix, or Use VOD" (March 6, 2017), available at: http://www.leichtmanresearch.com/press/030617release.pdf.

5   Jonathan Godfrey, Courtney Bernard, Nick Miller, ACT | The App Association, "State of the App Economy" (4th ed.) (January 2016), available at: https://actonline.org/wp-content/uploads/2016_State_of_App_Economy.pdf.

framework designed for now-defunct monopolies, analog technologies, and rigid divisions between service types based on
outmoded techno-functional distinctions. Multi-channel video
programming distributors (MVPDs), such as cable and direct
broadcast satellite providers (DBS), are subject to sector specific
regulations under Title VI of the Communications Act. Broadcast
TV and wireless service providers are subject to specific regulatory restraints under Title III. Wireline voice providers and other
providers offering "telecommunications services" are subject to
common carrier regulation under Title II. All of those services
may also be subject to regulation pursuant to the FCC's general
jurisdictional grant under Title I, including when the Commission deems regulation reasonably ancillary to the performance
of its mandated responsibilities. Additional platform-specific
rules apply to specific service segments, such as incumbent local
exchange carriers (ILECs) and local cable operators.

The Communications Act's legacy telecommunications regulatory regime is mostly premised on consumers receiving voice
telephone service from a monopolistic local exchange carrier
and a monopolistic long-distance carrier. Similarly, the existing
framework is premised on cable operators possessing a 90-plus
percent share of the video subscription market.

But those premises have been upended by market developments. For instance, wireless services have grown exponentially
over the past two decades. By year's end 2015, total wireless connections in the U.S. exceeded 374 million.[6] According to a Center
for Disease Control-National Center for Health Statistics survey,
50.8% of adults lived in households that are wireless-only.[7] Data

---

6   FCC, Implementation of Section 6002(b) of the Omnibus Budget Reconciliation
    Act of 1993; Annual Report and Analysis of Competitive Market Conditions
    With Respect to Mobile Wireless, Including Commercial Mobile Services,
    WT Docket No. 16-137, *Nineteenth Report* ("*Nineteenth Wireless Competition
    Report*"), at 8, ¶ 12 (September 23, 2016), available at: https://apps.fcc.gov/
    edocs_public/attachmatch/DA-16-1061A1.pdf.
7   Stephen J. Blumberg and Julian V. Lake, "Wireless Substitution: Early Release

collected by the National Telecommunications and Information Administration (NTIA) indicates that among all online households, 20% were wireless-only by 2015.[8] Also, as a result of entry by traditional cable operators into the voice market, 60 million consumers subscribed to Voice over Internet Protocol (VoIP) services by the middle of 2016.[9] Traditional telephone switched access lines plummeted to 62 million.[10] For consumers, the distinction between "local" and "long distance" calling is now an anachronism. Voice providers typically offer consumers a single flat-rate monthly bill or options for pre-paid bucket plans.

At the end of 2015, two nationwide DBS providers served about one third of the video subscription market.[11] By that same time, traditional telephone companies provided video services to 13.4% of MVPD subscribers.[12] Cable's market share dropped to 53.1% of MVPD subscribers.[13] However, total MVPD subscriptions have declined each year since 2013. Overall MVPD subscriptions dropped 1.1 million over the course of 2014 and 2015,[14] for

of Estimates From the National Health Interview Survey, July-December 2016," Division of Health Interview Statistics, National Center for Health Statistics (released May, 2017), available at: https://www.cdc.gov/nchs/data/nhis/earlyrelease/wireless201705.pdf.

8   Giulia McHenry, "Evolving Technologies Change the Nature of Internet Use," National Telecommunications & Information Administration (NTIA) (April 19, 2016), available at: http://www.ntia.doc.gov/blog/2016/evolving-technologies-change-nature-internet-use.

9   FCC (Wireline Competition Bureau), Voice Telephone Services: Status as of June 30, 2016 ("*Voice Telephone Services Report*"), at 2-3, 5 (April 2017), available at: https://apps.fcc.gov/edocs_public/attachmatch/DOC-344500A1.pdf.

10   *Voice Telephone Services Report*, at 2-3, 5.

11   FCC (Media Bureau), Annual Assessment of the Status of Competition in the Market for the Delivery of Video Programming, MB Docket No. 16-247, *Eighteenth Report* ("*Eighteenth Video Competition Report*"), at 7-8 ¶ 19 (released January 17, 2017), available at: http://transition.fcc.gov/Daily_Releases/Daily_Business/2017/db0117/DA-17-71A1.pdf.

12   *Eighteenth Video Competition Report*, at 7-8 ¶ 19.

13   *Eighteenth Video Competition Report*, at 7-8 ¶ 19.

14   *Eighteenth Video Competition Report*, at 29, ¶ 68.

a 2015 year-end total of 100.5 million MVPD subscriptions.[15] In the first quarter of 2017, the top six cable providers reportedly lost an additional 115,000 subscribers.[16] One analyst estimated that as many as 762,000 subscribers to video services cut the cord in the first quarter of 2017,[17] with future declines in video subscriptions widely expected.

Meanwhile, subscriptions to over-the-top (OTT) or online video delivery (OVD) services via fixed or mobile broadband connections outnumber all MVPD subscriptions combined. Reports indicate that by the second quarter of 2016, Hulu subscriptions in the U.S. reached 12 million.[18] By the first quarter of 2017, Amazon Prime subscriptions in the U.S. climbed to 80 million and Netflix subscriptions in the U.S. surpassed 50 million.[19] According to a survey from early 2017, 64% of TV households in the U.S. subscribed to Amazon Prime, Hulu, or Netflix.[20] Online video services are widely viewed by consumers on advanced devices that are unaffiliated with MVPDs. In 2014, an estimated

---

15   *Eighteenth Video Competition Report*, at 30, ¶ 68 (Table III.A.5).

16   Leichtman Research Group, Press Release: "Major Pay-TV Providers Lost About 410,000 Subscribers in 1Q 2017" (May 18, 2017), available at: http://www.leichtmanresearch.com/press/051817release.html.

17   Peter Kafka, "Another half-million Americans cut the cord last quarter," *Recode* (May 3, 2017), available at: https://www.recode.net/2017/5/3/15533136/cord-cutting-q1-half-million-tv-moffett; Leitchman Research Group, Press Release,

18   Hulu, Press Release: "Hulu Goes Bigger and Bolder at 2016 Upfront Presentation, Unveils +30% Growth in Subscribes, New Programming Deals and Ad Partnerships (May 4, 2016), available at: https://www.hulu.com/press/hulu-goes-bigger-and-bolder-at-2016-upfront-presentation-unveils-30-growth-in-subscribers-new-programming-deals-and-ad-partnerships/. *See also Eighteenth Video Competition Report*, at 75, ¶ 180.

19   Stephanie Pandolph and Jonathan Camhi, "Amazon Prime subscribers hit 80 million," *Business Insider* (April 27, 2017), available at: http://www.businessinsider.com/amazon-prime-subscribers-hit-80-million-2017-4; Tom Huddleston, Jr. "Netflix Has More U.S. Subscribers Than Cable TV," *Fortune* (June 15, 2017), available at: http://fortune.com/2017/06/15/netflix-more-subscribers-than-cable/.

20   Leichtman Research Group, "82% of U.S. TV Households Have a DVR, Netflix, or Use VOD," *supra*.

nearly 91 million U.S. households with broadband connections used an average of 7.3 Internet-connected devices for viewing video, including streaming media devices such as Roku and TiVo, smart TVs and Blu-ray players.[21] And in 2016, there were more than 460 million IP-enabled consumer owned devices supporting video apps.[22] Also, by the end of 2016, an estimated 15% of U.S. broadband households had antenna-only TV service, featuring programming broadcast over-the-air in HD.[23]

Next-generation wireless networks now make mobile video a reality for consumers with smartphones and tablets. As of December 2015, 95.9% of the U.S. population lived in census blocks served by three or more 4G LTE mobile service providers and 89.1% lived in census blocks served by four or more providers.[24] By the first quarter of 2016, 80% of all mobile subscribers had smartphone devices.[25]

In short, new entry, along with technological innovation, have made cross-platform competition for voice, video, and data services into a defining feature of today's digital communications marketplace. Competitors in once disparate market sectors now vie for consumers as their respective product and service offerings place increasing competitive pressures on each other.

Faced with an explosion of new product and service options for consumers and a competitive landscape drastically reshaped by convergence around IP-based services, the old communications regulatory order increasingly is under severe strain. But, historically, rather than use whatever powers it possesses under the existing statute to reduce legacy regulatory burdens, the FCC

---

21   *Eighteenth Video Competition Report*, at 81 ¶ 194 (based on SNL Kagan, "U.S. Connected Devices, 2010-2019" (September 23, 2015)).

22   *Eighteenth Video Competition Report*, at 81 ¶ 194.

23   Parks Associates, Press Release: "15% of U.S. broadband households have antenna-only TV service" (January 18, 2017), available at: http://www.parksassociates.com/blog/article/pr-01182017.

24   *Nineteenth Wireless Competition Report*, at 30-31 ¶ 39 (internal cite omitted).

25   *Nineteenth Wireless Competition Report*, at 85 ¶ 121 (internal cite omitted).

has continued to enforce rules designed to address monopolistic conditions in insular service markets. Rather than come to terms with technological and market changes, the Commission has tended towards efforts to preserve a powerful regulatory role for itself in the Digital Age. This is the most persuasive reason that Congress needs to update the Communications Act.

The FCC's Pro-Regulatory Defaults Result in Ongoing FCC Overreach

Thankfully, the FCC under Chairman Ajit Pai's leadership is in the process of reversing, in whole or in part, some of the most damaging regulatory excesses of Chairman Tom Wheeler's tenure. But this does not mean that a new statutory framework for communications services in the Digital Age is unnecessary. Under the legacy regulatory regime, to a significant extent, pervasive government restrictions are the presumptive mode of operation and can be resurrected even if they are eliminated by the Trump Administration FCC. This institutional presumption favoring regulation generally treats the necessity of government restrictions as a given that requires parties seeking relief from regulation to justify that relief.

It often has been observed that regulators have a built-in bias in favor of regulation. A presumptive regulatory approach leaves little to counteract that bias. More specifically, a default presumption favoring regulation prolongs bureaucratic controls over markets long after agency rules have outlived their reason for being. The FCC's pro-regulatory presumptive outlook has served as an analytical roadblock to long overdue deregulatory action. Elimination of outdated legacy regulations is sorely needed to relieve providers of unnecessary costs and remove barriers to investment in next-generation broadband networks.

In fact, the case for clean-slate reform is bolstered by the FCC's regrettable history of neglect and obstruction when it comes to the actual deregulatory tools that Congress authorized the Commission to employ. Section 10 of the Communications Act provides that the Commission "shall forbear" from applying any regulation or statutory provision if it determines enforcement of such requirement "is not necessary" to ensure that telecommunications carriers' charges and practices are reasonable and "not necessary for the protection of consumers."[26] And Section 11 requires periodic reviews of Commission regulations in order to determine "whether any such regulation is no longer in the public interest as a result of meaningful economic competition between providers of such service."[27] The Commission has used these obviously deregulatory tools only on rare occasions, and even then only reluctantly and after much delay.

As will be discussed further below, the FCC's default preference for regulation is also reflected in its assertion of public utility-style and other legacy-type controls over advanced broadband technologies in today's competitive markets.

Broad Agency Discretion and Judicial Deference Enable FCC Overreach

Along with a presumptively pro-regulatory outlook, the FCC possesses excessively broad discretion in adopting and enforcing regulations. In over 100 instances, the Communications Act confers on the Commission authority to regulate in the "public interest"[28] – an indeterminate standard that enables the Commission to exercise almost unfettered regulatory power. Such broad discretion facilitates the Commission's continuation of its legacy regulatory apparatus despite dynamic market changes that have upended the

---

26   47 U.S.C. § 160(a)(1)-(3).

27   47 U.S.C. § 161(a)(1)-(2).

28   *See* Randolph J. May, "The Public Interest Standard: Is It Too Indeterminate to Be Constitutional?", 53 *Fed. Comm. L.J.* 427, at 456-67 (2001) (listing provisions in the Communications Act that pertain to the public interest standard).

rationale for its outdated rules. The Commission's equally unjustifiable extension of public utility and other legacy-type regulations to advanced digital communications technologies, including high-speed broadband, is also facilitated by agency reliance on what it claims as broad discretion.

Exceedingly broad judicial deference to agency exercises of power similarly enables the FCC to give protracted life to outdated regulations that offer no apparent benefit to consumers in today's competitive digital communications ecosystem. Like other federal agencies, the FCC is the beneficiary of "*Chevron* deference." Under the Supreme Court's *Chevron* doctrine, when statutory language is silent or ambiguous regarding an issue, courts give an agency's own interpretation of that language "controlling weight" and thereby almost always accede to it.[29] The narrow exception is where courts conclusively determine that the agency's interpretation is unreasonable.[30] Similarly, under the *Auer/Seminole Rock* doctrine, courts will enforce an agency's interpretation of its own regulation unless it is "plainly erroneous or inconsistent with the regulation."[31] This high degree of judicial deference makes it difficult to hold the Commission accountable to the rule of law.

The Administrative Procedure Act (APA) also embodies a highly deferential judicial standard for reviewing agency action. Under the APA, courts reviewing the factual assertions and rationales for new regulations invalidate agency actions only when they are conclusively found to be "arbitrary, capricious, and abuse of discretion, or otherwise not in accordance with law."[32] Further, courts reviewing agency decisions made after formal adjudications and rulemakings will uphold such decisions merely when "substantial evidence" supports them.[33] "Substantial evidence" is

29   *Chevron U.S.A. Inc. v. Natural Res. Def. Council*, 467 U.S. 837, 844 (1984).

30   *Chevron*, 467 U.S. 837 at 844.

31   *See Bowles v. Seminole Rock & Sand Co.*, 325 U.S. 410, 414 (1945); *Auer v. Robbins*, 519 U.S. 452, 461 (1997).

32   5 U.S.C. § 706(2)(A).

33   5 U.S.C. § 706(2)(E); 5 U.S.C. § 556(d).

a low bar standard requiring only "more than a mere scintilla" of "relevant evidence as a reasonable mind might accept as adequate to support a conclusion."[34] Functionally, the APA's arbitrary and capricious standard and its substantial evidence standard are equivalents. Both APA standards provide FCC actions with significant insulation from judicial checks and balances.[35]

The Supreme Court and federal appeals courts have provided occasional checks against FCC abuses of regulatory power. Notably, the Commission's first two attempts to regulate the Internet through so-called net neutrality restrictions on broadband network practices were struck down by the U.S. Court of Appeals for the D.C. Circuit in 2010 and 2014.[36] Those court decisions and others,[37] to some extent, rebuffed FCC power grabs and reasserted the rule of law. Yet the tide of overly expansive and arbitrary regulation of digital communications services by the Commission has continued largely unchecked.

The Supreme Court and lower courts have recognized that protected free speech interests warrant intermediate scrutiny of regulatory restrictions on communications services when First Amendment claims are involved.[38] But on the whole, where cases

---

34   *Consolidated Edison Co. v. NLRB*, 305 U.S. 197 (1938); *Metropolitan Stevedore Co. v. Rambo*, 521 U.S. 121, 149 (1997); *HTH Corp. v. NLRB*, 823 F.3d 668 (D.C. Cir. 2016).

35   *See, e.g., U.S. Telecom Ass'n. v. FCC*, 825 F.3d 674 (D.C. Cir. 2016); *Qwest Corp. v. FCC*, 689 F.3d 1214 (10th Cir. 2012).

36   *Comcast Corp. v. FCC*, 600 F.3d 642 (D.C. Cir. 2010); *Verizon v. FCC*, 740 F.3d 623 (D.C. Cir. 2014).

37   *U.S. Telecom Ass'n. v. FCC*, 359 F.3d 554, 566 (D.C. Cir. 2004); *Verizon Telephone Companies v. FCC*, 570 F.3d 294 (D.C. Cir. 2009); *Comcast Corp. v. FCC*, 579 F.3d 1 (D.C. Cir. 2009); *EchoStar v. FCC*, 704 F.3d 992 (D.C. Cir. 2013); *Comcast Cable Comm., LLC, v. FCC*, 717 F.3d 982 (D.C. Cir. 2013), order on remand affirmed by *The Tennis Channel, Inc. v. FCC*, 827 F.3d 137 (D.C. Cir. 2016).

38   *See, e.g., Turner Broadcast Systems, Inc. v. FCC* ("*Turner I*"), 512 U.S. 622, 636 (1994); *Turner Broadcasting Systems, Inc. v. FCC* ("*Turner II*"), 520 U.S. 180 (1997); *Time Warner Cable v. FCC*, 729 F.3d 137, 154-167 (2d Cir. 2013); *Time Warner Cable v. Hudson*, 667 F.3d 630, 638-642 (5th Cir. 2012); *Ill. Bell Tel. Co. v. Village of Itasca*, 503 F. Supp. 2d 928, 947-949 (N.D. Ill. 2007); *Comcast*

involving FCC infringement of protected constitutional speech rights have come before courts, those cases have been decided on other grounds, with First Amendment considerations left unaddressed or only addressed tangentially.[39]

As will be explained in this book, a new Digital Age Communications Act is needed to establish an evidentiary presumption against regulation in the absence of clear and convincing evidence to the contrary. FCC regulatory actions should be subject to a competition-based standard that requires the Commission to apply sound economic analysis in assessing market effects and the impact on overall consumer welfare. Under a statute incorporating such a deregulatory presumption and competition-based standard, courts will be less inclined to find that regulatory overreaches pass muster. While it is ultimately up to the Supreme Court to develop more protective First Amendment free speech jurisprudence for communications media in the Digital Age, adoption of a statutory standard tied to actual competitive market conditions may prompt the courts to scrutinize more rigorously regulatory restrictions that impinge upon the free speech rights of communications service providers. At the very least, Congress can provide significant incidental protections to free speech rights by replacing the FCC's broad discretionary authority for regulating communications services and instead requiring that its regulatory actions satisfy the strictures of a competition-based standard.

Guiding Principles for a Modernized Communications Act

We have described the significant problems with our current communications laws and regulations. Now we will set forth guiding

Cablevision of Broward County, Inc. v. Broward County, 124 F. Supp. 2d 685, 690-698 (S.D. Fla. 2000).

39 *See, e.g., Fox v. FCC*, 556 U.S. 502, 530 (Thomas, J., concurring); *Comcast Corp. v. FCC*, 579 F.3d at 9. *See also, e.g., Comcast Cable Commc'ns, LLC v. FCC*, 717 F.3d 982, 994 (D.C. Cir. 2013) (Kavanaugh, J., concurring); *Comcast v. Cablevision v. FCC*, 597 F.3d 1306, 1318 (D.C. Cir. 2010) (Kavanaugh, J., dissenting).

principles for replacing the current statute with a modernized Digital Age Communications Act and explain the reasoning behind those principles.

In important respects, the clean-slate approach recommended in this book owes to the 2005 Digital Age Communications Act (DACA) proposal that was produced by working groups of scholars under the auspices of the then-Progress and Freedom Foundation.[40] Several contributors to the papers that comprise the chapters in this monograph were participants in the DACA working groups. Undoubtedly, the technological and competitive landscape has undergone significant changes over the last dozen years. While building upon the same market-based principles of the DACA proposal, this book offers more up-to-date analysis and recommendations for bringing communications policy into the Digital Age.

The occasion for the chapters comprising this book was the review of the Communications Act conducted by the U.S. House Committee on Energy and Commerce and its Communications and Technology Subcommittee – a project often simply referred to as "#CommActUpdate." Between January - December 2014, the Committee issued a series of White Papers aimed at seeking public input regarding whether and how federal communications law should be updated to address the competitive landscape of the 21st Century.[41] Each of the six White Papers released by the House Committee addressed a specific area of communications policy and sought public comment on particular questions posed. The chapters are the responses submitted by scholars affiliated with the Free State Foundation that are publicly available on the House Committee website. They are reproduced here in their

---

40 Randolph J. May, James B. Speta, *et al.*, "Digital Age Communications Act: Proposal of the Regulatory Framework Working Group: Release 1.0," Progress and Freedom Foundation (June 2005), available at: http://www.pff.org/issues-pubs/other/050617regframework.pdf.

41 U.S. House, Committee on Energy and Commerce, "#CommActUpdate," available at: https://energycommerce.house.gov/commactupdate.

original form, save for only minor non-substantive edits. We have prefaced each chapter with our own short summary of the House Committee White Paper that is the subject of the response and included relevant website information for readers desiring further background.

Aside from the monograph's editors – Free State Foundation President Randolph J. May and Senior Fellow Seth L. Cooper – all of the contributing authors of the numbered chapters are full-time academics who are members of the Free State Foundation's prestigious Board of Academic Advisors. Needless to say, the contributors are distinguished in their own rights. The authors collectively draw on their wealth of expertise and experience in setting forth principles and ideas that should shape reform of federal communications law.

DISCLAIMER: Not all of the contributors to the individual papers agreed on all points, of course, and they specifically disclaim doing so in each paper. Nevertheless, they agreed on the importance of the project. And they agreed, generally, on the need for meaningful reform of the Communications Act in the direction of less reliance on regulation and more reliance on marketplace competition to protect consumers from harm and to enhance their welfare. Nor should the title of this book be misconstrued to mean that it was published in any official capacity or that its substantive recommendations represent the views of the U.S. House Committee on Energy and Commerce.

Overview of the Chapters

Chapter 1 is a response to the Committee's White Paper "Modernizing the Communications Act." This chapter sets forth major guiding principles for reforming federal communications law and policy. In particular, it calls upon Congress to adopt a clean slate approach that would replace the existing regulatory structure with an entirely new Digital Age Communications Act or similar legislation that reflects today's marketplace realities. While almost all entities providing communications services will be subject to the FCC's jurisdiction under the new act, the Commission's authority to actually regulate such entities should be tied closely to a competition standard. In other words, the Commission's authority to regulate should be grounded in evidentiary findings regarding market failure and consumer harm, not vague allusions to the "public interest." Indeed, except perhaps with a few exceptions relating to public safety or the like, the Communications Act's present ubiquitous delegations to the Commission to act in the "public interest" should be cut back severely.

The Communication Act's present "stovepipe" or "silo" regulatory regime that applies disparate regulatory requirements to comparable services based on now outdated techno-functional constructs should be eliminated. The FCC's power to impose broad anticipatory rules regarding marketplace conduct should be curtailed. Commission rules should sunset after fixed year terms absent a strong evidentiary showing that they remain necessary to address existing harms and therefore ought to be extended. Under a new act, the primary means for the Commission to address market conduct or providers' practices should be case-by-case adjudication of specific complaints alleging that specific conduct or practices violate a particular law or regulation in a way that causes harm to consumers or competition. Chapter 1 also calls for FCC jurisdiction over privacy and data security to be transferred to the Federal Trade Commission (FTC), and for state authority over digital communications service providers to be circumscribed by the new act.

Chapter 2 is a response to the Committee's White Paper "Modernizing U.S. Spectrum Policy." It sets forth general concepts for implementation of a new federal policy that reflects advances in spectrum technology, including new uses, convergence of applications, and skyrocketing consumer demand. The existing administrative fiat model for spectrum licensing – including FCC allocation of particular frequency bands for pre-specified service uses and technical parameters – is economically inefficient and should be discarded. In its place, Congress should establish a robust, flexible market-based system in which spectrum rights can be freely traded, and with little administrative control by the Commission. Unencumbered auctions should supply the means for initial assignment of spectrum rights, with free trade arrangements governing the secondary market for spectrum rights. Under a market-based model, the Commission would retain a more limited role, similar to that of a rights clearinghouse operator or registrar of deeds, with modest zoning-like powers to address technical coordination problems. The Commission also would retain authority to address special spectrum-related needs, such as assuring public safety.

Chapter 3 is a response to the Committee's White Paper "Competition Policy and the Role of the Federal Communications Commission." FCC attempts to "manage competition" among various service sectors through resale or sharing mandates and other command-and-control techniques such as rate controls fail to account for cross-platform competition and should end. Enhancing consumer welfare in the Digital Age requires that Commission policy implementation actually take into account the existence of ever-increasing competition among platforms that were once labeled as disparate services simply because of legacy constructs having to do with operational features and functions or technologies employed. Those constructs are no longer relevant because today's competing platforms generally comprise the same relevant market from a consumer's perspective.

This chapter lays out a consistent, principled competition policy framework premised on facilitating free entry and exit as the basic rule, qualified by FCC application of targeted *ex post* remedies to address anticompetitive conduct concerns – rather than broad *ex ante* regulations of specific services. The touchstone of the Commission's competition standard should be consumer welfare, rather than competitor welfare or the indeterminate "public interest." This new standard should rely heavily on anti-trust-like jurisprudence grounded in sound economic analysis to assess claims of consumer harm or harm to competition. Under this standard, where concerns about market conduct arise based on individual complaints containing specific allegations of abuse, the Commission would rely on focused adjudicatory proceedings. Critically, the party seeking regulatory remedies should bear the burden of providing clear and convincing evidence of violations of law or regulation tied to market power abuse and consumer harm.

Chapter 4 is a response to the Committee's White Paper "Network Interconnection." It calls for the FCC to assume a much less interventionist role in overseeing the interconnection of privately-operated networks that comprise the nation's 21st Century digital communications infrastructure than the role that the Commission has served regarding interconnection of the 20th Century public switched telephone network. Under a new act, the Commission's authority regarding interconnection should be tied to a presumption that interconnection agreements between IP-based networks shall be the subject of voluntary negotiation. Regulatory intervention by the Commission should only take place upon an evidentiary finding that denial of interconnection poses a substantial, non-transitory risk to consumer welfare that competitive market forces cannot sufficiently address. In such circumstances, the Commission should be authorized to use dispute resolution mechanisms such as mediation or arbitration to address the impasse and ultimately facilitate agreement between parties on terms for interconnection.

Chapter 5 is a response to the Committee's White Paper "Universal Service Policy and the Role of the Federal Communications Commission." Universal service is the federal policy – or at least the policy goal – for ensuring that all Americans have access to basic communications services. It is a subsidy program, the overall budget for which now exceeds $8.3 billion annually.[42] Program funding is ultimately drawn from consumers of interstate and international voice services in the form of Universal Service Fund (USF) surcharges included on monthly bills. A new act should refocus universal service policy on assisting consumers who cannot afford broadband service, using market-based initiatives that empower consumers, such as means-tested, individualized vouchers.

Also, USF support for infrastructure build-outs should be limited to areas that are unserved so as not to support duplicate providers. Build-out support generally should be accomplished through reverse auctions that grant targeted subsidies to those bidding service providers that offer to provide the specified service at the lowest rate. Finally, funding for universal service should be shifted to appropriations from Congress from the general treasury. This approach offers greater transparency, accountability, and fiscal control than currently exists with the application of a stealth "USF surcharge" – which in practical effect is a tax imposed on consumers – applied to consumers' voice service bills.

Chapter 6 is a response to the Committee's White Paper "Regulation of the Market for Video Content and Distribution." Given the dramatic changes in the video landscape, including the proliferation of video delivery service options through new online offerings and devices, the legacy regulations governing video services need to be largely eliminated. Consumers now enjoy an abundance of choices when it comes to consumption of video. Consistent with the recommendations contained in prior chapters, Chapter 6 calls for discarding the disparate regulatory regimes now governing comparable video service offerings and also for jettisoning, for the most part, the indeterminate "public interest" standard that applies to much regulatory activity in this

---

42   Universal Service Administrative Company (USAC), "Building the Foundation: 2015 Annual Report," at 40, available at: http://www.usac.org/_res/documents/about/pdf/annual-reports/usac-annual-report-interactive-2015.pdf.

area. As with other digital communications services, in a new act regulation of video services should be dependent on clear and convincing evidence of market failure and consumer harm before government intervention in the market is permitted. Here again, a new act should make case-by-case adjudication of specific allegations of market power and consumer harm the primary mode for FCC enforcement. Since selection and placement of video content and decisions regarding bundling of programming by service providers involve editorial decisions that receive First Amendment protection, this chapter's analysis and recommendations are informed and bolstered by reliance on constitutional free speech protections.

By no means do these chapters seek to address every area of communications law in need of reform, and certainly not in minute detail. Nor do these chapters address other matters such as state and local government restrictions on wireless infrastructure construction and modification or taxation and fees imposed on telecommunications services. Undoubtedly, a modernized Communications Act that embodies a clean-slate approach must contain specific dates for sunsetting certain Communications Act provisions, including provisions relating, for example, to the public switched telephone network, the unbundling/forced sharing mandates, and outdated media ownership restrictions. Of course, there will be a need to spell out transitional periods in many instances to avoid or minimize disruptions. However, any necessary reforms regarding those matters should also be addressed in light of the core precepts set out in the chapters.

A short conclusion, prepared by this book's editors, summarizes the guiding principles for modernizing the Communications Act that are presented in the chapters. An appendix addressing needed agency process reforms is also included. The appendix consists of Congressional testimony of Randolph J. May in support of the proposed Federal Communications Commission Process Reform Act of 2013 as well as other proposed reforms. Similar legislation subsequently passed the House of Representatives in January 2017.[43] The reforms addressed in the appendix should be included in an appropriate manner in a new act.

Recent Developments in Today's Competitive Marketplace Bolster the Case for Modernizing of the Communications Act

In the time since the chapters were prepared for submission to the House Energy and Commerce Committee, their underlying free market-oriented premises have been bolstered by new data and market developments. Developments from 2014 through 2016 have made even more pressing the need for a "clean slate" approach to modernizing the Communications Act.

Data contained in the FCC's *Nineteenth Wireless Competition Report* (2016), *Eighteenth Video Competition Report* (2016), and *2016 Broadband Progress Report* reconfirm the dynamic, competitive trends described throughout the chapters.[44] Because the chapters cite to data collected in prior reports, readers may wish to consult the most recent iterations for updated statistics and estimates. Of course, even the latest reports released by the Commission are typically a year or more out-of-date concerning market developments.

---

43   *See* U.S. Congress. House. *Federal Communications Commission Process Reform Act of 2017*. 115th Cong. 1st sess. H.R. 290, available at: https://www.congress.gov/bill/115th-congress/house-bill/290; U.S. Congress. House. *Federal Communications Commission Consolidated Reporting Act of 2017*. 115th Cong. 1st sess. H.R. 599, available at: https://www.congress.gov/bill/115th-congress/house-bill/599.

44   *Nineteenth Wireless Competition Report, supra; Eighteenth Video Competition Report, supra;* FCC, Inquiry Concerning the Deployment of Advanced Telecommunications Capability to All Americans in a Reasonable and Timely Fashion, and Possible Steps to Accelerate Such Deployment Pursuant to Section 706 of the Telecommunications Act of 1996, as Amended by the Broadband Data Improvement Act, GN Docket No. 15-191, *2016 Broadband Progress Report* (released January 29, 2016), available at: https://apps.fcc.gov/edocs_public/attachmatch/FCC-16-6A1.pdf.

Recent market merger activity also reinforces the critical role of cross-platform competition and convergence of IP-based services in the Digital Age. In July 2015, the Department of Justice (DOJ) and the FCC allowed the AT&T/DirecTV merger.[45] Also, in October 2016, AT&T announced an agreement to acquire media and entertainment company Time Warner, Inc.[46] The Charter Communications/TimeWarner Cable/Bright House Networks merger received permission from both reviewing federal agencies in April and May of 2016.[47] Verizon Communications acquired AOL in 2015, and its acquisition of XO Communications was approved by the FCC in November 2016.[48] In July 2016, Verizon announced it reached an agreement to acquire Yahoo's operating business,[49] and the merger consummated in June 2017. CenturyLink announced a proposed merger with Level 3 Communications in October 2016, with the transaction expected to close sometime in 2017.[50] And in May 2017, cable operators Comcast

45 FCC, Applications of AT&T Inc. and DIRECTV For Consent to Assign Transfer Control of Licenses and Authorizations, MB Docket No. 14-90, Memorandum Report and Order ("AT&T/DirecTV Order"), (released July 28, 2015), available at: https://apps.fcc.gov/edocs_public/attachmatch/FCC-15-94A1.pdf.

46 AT&T, Press Release: AT&T to Acquire Time Warner" (October 22, 2016), available at: http://about.att.com/story/att_to_acquire_time_warner.html.

47 FCC, Applications of Charter Communications, Inc., Time Warner Cable, Inc., and Advance/Newhouse Partnership For Consent to Assign or Transfer Control of Licenses and Authorizations, Memorandum Opinion and Order ("*Charter/ Time Warner Cable Order*"), MB Docket No. 15-149, (released May 10, 2017), available at: https://apps.fcc.gov/edocs_public/attachmatch/FCC-16-59A1.pdf.

48 FCC (Wireline Competition Bureau), Applications of XO Holdings and Verizon Communications Inc. For Consent to Transfer Control of Licenses and Authorizations, WC Docket No. 16-70, Memorandum Opinion and Order ("XO/Verizon Order"), (released November 16, 2016), available at: https://apps.fcc.gov/edocs_public/attachmatch/DA-16-1281A1.pdf.

49 Verizon, Press Release: "Verizon to acquire Yahoo's operating business" (July 25, 2016), available at: http://www.verizon.com/about/news/verizon-acquire-yahoos-operating-business.

50 CenturyLink, Press Release: "CenturyLink to acquire Level 3 Communications" (October 31, 2016), available at: http://news.centurylink.com/news/centurylink-to-acquire-level-3-communications.

and Charter Communications announced an agreement to co-operate in developing and launching competitive mobile wireless services.[51]

Further, broadband and video service providers have acquired interests in or started up several new OTT or OVD ventures.[52] Online video delivery – including video delivery over wireless broadband systems – lies at the heart of the cross-platform competitive and convergent market. Meanwhile, speculation continues about future mergers between cable and wireless entities. Both the reality and potentialities of combinations among providers once regarded as serving entirely separate market sectors points unmistakably to the dynamic, convergent nature of today's digital communications services market.

The FCC Has Forced the Legacy Regulatory Regime on Today's Competitive Digital Marketplace

The need for a Communications Act overhaul has become even more urgent in light of the FCC's more aggressive imposition of new regulations over broadband and video services during the tenure of Chairman Tom Wheeler. Through several major actions taken between 2014 and 2016, along with some earlier actions, the Obama Administration's FCC revealed a persistent preference for regulatory interference with today's dynamic broadband marketplace. Absent any meaningful, much less convincing, evidence of market power harms and in the face of dramatic advances in technology and competition across platforms,[53] the Commission

---

51   Comcast, Press Release: "Comcast, Charter to Explore Operational Efficiencies to Speed Entry into Wireless Market" (May 8, 2017), available at: http://corporate.comcast.com/news-information/news-feed/comcast-charter-wireless-efficiencies.

52   *See, e.g., Eighteenth Video Competition Report*, at 65, ¶ 154. *See also id.* at 65-66, ¶ 155.

53   FCC, Protecting and Promoting the Open Internet, GN Docket No. 14-28, Report and Order on Remand, Declaratory Ruling, and Order ("*Open Internet*

deliberately subjected the Digital Age's competitive communications marketplace to legacy regulatory constraints.

In the Telecommunications Act of 1996, Congress paid considerable attention to local and long-distance voice services markets, while devoting scarce attention to the Internet or broadband services. Yet, the FCC asserted its authority over broadband Internet access services and other services through questionable interpretations of federal law. In its *Open Internet Order* (2015), the Commission staked broad jurisdictional authority over the management of broadband networks on Section 706 of the 1996 Act – a provision wherein Congress exhorts the Commission to use a series of deregulatory measures to remove barriers to investment.[54] From among Section 706's short list of methods for reforming or removing burdens on incumbent local exchange carriers, the Commission claimed to find a grant of authority to impose public utility-style regulation on the information technology medium and economic driver of the 21st Century – the Internet.

And the 2015 *Open Internet Order* also imposed public utility-style regulation on broadband Internet access services through equally dubious claims to authority under Title II of the Communications Act. The *Open Internet Order* force-fit broadband Internet access services within the class of Title II "telecommunications services."[55] At its core, Title II is a regime for imposing rate regulation and non-discrimination mandates. The Commission's decision to reclassify broadband Internet access services was therefore a decision to assume intrusive regulatory oversight over rates and network management practices. Its Title II reclassification decision reversed prior Commission determinations that regarded broadband Internet access services as

---

Order"), at 6, ¶¶ 11 fn.12 (released March 12, 2015), available at: https://apps.fcc.gov/edocs_public/attachmatch/FCC-15-24A1.pdf.

54   *Open Internet Order*, at 121-124, ¶¶ 275-282.

55   *Open Internet Order*, at 157-177, ¶¶ 355-387.

lightly regulated Title I "information services."[56] The prior Title I policy, which was upheld by the Supreme Court in *NCTA v. Brand X* (2005),[57] spurred tremendous entrepreneurial investment in broadband infrastructure. Between 2005 and 2014, total wireline broadband capital investment equals approximately $695 billion.[58]

As a means for subjecting wireless to Title II, the *Open Internet Order* even redefined the public switched telephone network to include mobile broadband Internet access services.[59] This Title II reclassification decision subjected advanced wireline and wireless broadband Internet access services, including 4G Long Term Evolution (LTE) networks and gigabit speed fiber-optic networks, to seemingly open-ended public utility-style regulatory authority.

To further its pro-regulatory inclinations, the Obama Administration FCC created its own *ad hoc* version of Title II regulation for the Internet. It did so by twisting yet another

---

56 *See, e.g.,* FCC, Inquiry Concerning High-Speed Access to the Internet Over Cable and Other Facilities; Internet Over Cable Declaratory Ruling, GN Docket No. 00-85; Appropriate Treatment for Broadband Access to the Internet Over Cable Facilities, CS Docket No. 02-52, Declaratory Ruling and Notice of Proposed Rulemaking (*"Cable Modem Declaratory Order"*)(released March 15, 2002), available at: https://apps.fcc.gov/edocs_public/attachmatch/FCC-02-77A1.pdf, affirmed by *National Cable & Telecomms. Ass'n. v. Brand X Internet Services*, 545 U.S. 967 (2005); FCC, Appropriate Framework for Broadband Access to the Internet over Wireline Facilities, Report and Order and Notice of Proposed Rulemaking (*"Wireline Broadband Internet Access Services Order"*), CC Docket 02-33, *et al.,* (released September 23, 2005), available at: https://apps.fcc.gov/edocs_public/attachmatch/FCC-05-150A1.pdf, affirmed by *Time Warner Telecom, Inc. v. FCC*, 507 F.3d 205 (3d Cir. 2007); FCC, Appropriate Treatment for Broadband Access to the Internet Over Wireless Networks, WT Docket No. 07-53, Declaratory Ruling (*"Wireless Broadband Order"*) (released March 23, 2007), available at: https://apps.fcc.gov/edocs_public/attachmatch/FCC-07-30A1.pdf.

57 545 U.S. 967 (2005).

58 *See* Patrick Brogan, "Broadband Investment Gains Continued in 2014: Research Brief," USTelecom, at 2 (Chart 2) (July 24, 2015), available at: https://www.ustelecom.org/sites/default/files/documents/Investment-2014-Research-Brief-July-2015.pdf.

59 *Open Internet Order*, at 178-190, ¶¶ 388-408.

statutory provision: Section 10.[60] That section grants the Commission unique authority to forbear from applying telecommunications regulations that are outdated and unnecessary for protecting consumers. The primary purpose of Section 10 is deregulatory, that is, to give the Commission a mechanism for relieving incumbent telephone companies from costly compliance burdens with legacy telecommunications regulations if and when competitive conditions emerged in voice services markets. But in its *Open Internet Order*, the Commission invoked Section 10 to forbear from applying several Title II provisions to broadband services.[61] In the FCC's wishful thinking, applying Section 10 in a "just-right" manner would "implement the strongest possible rules to protect net neutrality" but would also somehow protect the Internet from being flooded by Title II public utility regulations.[62] Unfortunately, in *US Telecom v. FCC* (2016), the U.S. Court of Appeals for the District of Columbia Circuit upheld the FCC's *Open Internet Order*, including the Commission's unwise Title II reclassification decision and dubious usage of Section 10.[63]

Also unfortunate was the FCC's imposition of public utility-style regulation on broadband services when consensus reasonably could have been reached on a more moderate regulatory policy based on Title I. Transparency requirements for broadband ISPs are largely uncontroversial. Agreement quite possibly could have been achieved among the Commissioners regarding rules to prohibit ISPs from blocking as well as throttling or degrading access to legal content on the Internet. Yet on a 3-2 partisan vote, the

---

60   47 U.S.C. §160(a)(1)-(3).

61   *Open Internet Order*, at 204-267 ¶¶ 434-542.

62   The White House, Office of the Press Secretary, Press Release: "Statement by the President on Net Neutrality" (November 10, 2014), available at: https://obamawhitehouse.archives.gov/the-press-office/2014/11/10/ statement-president-net-neutrality.

63   *United States Telecommunications Association v. FCC, 825* F.3d 674 (D.C. Cir. 2016), rehearing *en banc* denied, 855 F.3d 381 (D.C. Cir. 2017).

Commission instead pushed through absolutist rules concerning those practices based on Title II.[64] The Commission also adopted a bright-line rule banning all "paid prioritization" arrangements.[65]

Indeed, the *Open Internet Order*'s ban on charging edge content providers for high-volume delivery or higher service quality prompted former FCC Chief Economist Tim Brennan to describe the order as an "economics-free zone."[66] It is common knowledge that certain online content or edge providers generate outsized amounts of Internet traffic in relation to most other providers. If broadband ISPs can charge those edge providers fees that reflect at least some of the outsized usage they generate – and the associated costs imposed on ISPs' networks – lighter users would not be forced, in effect, to subsidize those entities that generate much heavier usage of ISPs' facilities. In other words, consumers – quite reasonably – may prefer that some forms of high-volume traffic be prioritized as long as those who benefit pay more for such prioritization, thereby reducing the price low-volume traffic pays. But the FCC's mandate prevents all experimentation with such prioritization plans, even though they are common in many other delivery markets.

The FCC's newly assumed authority over broadband was rendered more injurious by the adoption of its "general conduct" or "no-unreasonable interference/disadvantage" standard. In addition to subjecting broadband service providers to what are termed "Clear, Bright Line Rules,"[67] the *Open Internet Order* also subjects providers to what the Obama Administration Commission unabashedly called a "catch-all" provision. This "general

---

64   *Open Internet Order*, at 69-82, ¶¶ 154-184; *id.* at 45-53, ¶¶ 104-124.

65   *Open Internet Order*, at 53-58, ¶¶ 125-132.

66   *See* Tim Brennan, "Is the Open Internet Order an 'Economics-Free Zone'? *Perspectives from FSF Scholars*, Vol. 11, No. 22 (June 28, 2016), available at: http://www.freestatefoundation.org/images/Is_the_Open_Internet_Order_an_Economics_Free_Zone_062816.pdf.

67   *Open Internet Order*, at 47-58, ¶¶ 110-132.

conduct" standard or "catch-all" declares that broadband service providers:

> [S]hall not unreasonably interfere with or unreasonably disadvantage (i) end users' ability to select, access, and use broadband Internet access service or the lawful Internet content, applications, services, or devices of their choice, or (ii) edge providers' ability to make lawful content, applications, services, or devices available to end users. Reasonable network management shall not be considered a violation of this rule.[68]

The order indicated that the Commission would apply the catch-all standard on a case-by-case basis, taking into account the "totality of the circumstances," guided by a list of seven equally vague factors plus the promise that the list is "non-exhaustive."[69]

The misguided nature of the this "catch-all" provision was reflected in the FCC's investigation into "free data" or "zero-rated" plans offered to consumers by mobile broadband providers. Free data or zero-rated plans, in one way or another, allow subscribers to avoid data usage charges when accessing certain websites, such as music or video sites. Although such plans differ in their details, in general they offer price-sensitive consumers a low-cost alternative for accessing data they otherwise would forego. Free data plans are therefore pro-adoption and pro-consumer, benefitting new and low-income subscribers in particular. In other words, they are a means of addressing the oft-discussed digital divide.

Although the *Open Internet Order* declared that free data plans were permissible, the order indicated those plans would be

---

68   *Open Internet Order*, at 59-60, ¶¶ 133-137.
69   *Open Internet Order*, at 61-64, ¶¶ 138-145.

subject to its "general conduct" standard.[70] Shortly after the order was issued, the FCC's Wireless Bureau conducted an investigation of free data plans offered by mobile providers that dragged out for more than a year. With the approval of Chairman Wheeler, the Wireless Bureau finally issued a report on its investigation in early 2017. It purported to find that free data programs offered by AT&T and Verizon Wireless were contrary to Open Internet principles.[71] Relying on its "catch-all" provision, the FCC reached a baffling anti-consumer conclusion that mobile service plans allowing access to popular content without incurring extra data charges are actually harmful. Shortly thereafter, and with the approval of Chairman Pai, the Wireless Bureau promptly rescinded the report.[72]

The *Open Internet Order* has provided no discernable benefits to consumers. Meanwhile, adverse consequences of the FCC's aggressive assertion of authority and its generation of regulatory uncertainty in the marketplace already may have revealed themselves. According to one industry report, investment by broadband providers declined $1 billion in 2015 compared to the year before.[73] And according to another report, AT&T, Verizon, Comcast, and Time Warner Cable – all ranked in the top 25 U.S. companies for capital investment – "cut U.S. capital spending by

---

70   *Open Internet Order*, at 66-69, ¶¶ 151-153.

71   FCC (Wireless Telecommunications Bureau) "Wireless Telecommunications Bureau Report: Policy Review of Mobile Broadband Operators' Sponsored Data Offerings for Zero-Rated Content and Services" (released January 11, 2017), available at: http://transition.fcc.gov/Daily_Releases/Daily_Business/2017/db0111/DOC-342987A1.pdf.

72   FCC (Wireless Telecommunications Bureau), Wireless Telecommunications Bureau Report: Policy Review of Mobile Broadband Operators' Sponsored Data Offerings for Zero-Rated Content and Services, Order (released February 3, 2017) (rescinding report), available at: http://transition.fcc.gov/Daily_Releases/Daily_Business/2017/db0203/DA-17-127A1.pdf.

73   Patrick Brogan, "Broadband Investment Ticked Down in 2015: Research Brief," USTelecom, at 1-2 (December 14, 2016), available at: https://www.ustelecom.org/sites/default/files/Broadband%20Investment%20Down%20in%202015.pdf.

1.3 percent in 2015 as compared to 2014."[74] The *Nineteenth Wireless Competition Report* even observed that 2015 wireless service provider incremental spending on mobile broadband totaled $30.9 billion, "which is a decline of approximately 3.2 percent from the $31.9 billion invested in 2014."[75] For the years 2015 and 2016, Free State Foundation Research Fellow Michael Horney estimated foregone investment of $5.6 billion due to Title II regulation of broadband.[76] These apparent declines in broadband capital investment are the first since the Great Recession of 2009 and the telecom/dot-com crash of 2001.[77] While such declines may not be solely attributable to the *Open Internet Order*, it is likely that Title II regulation at least contributed to them.

Through its *Open Internet Order*, the FCC also purported to stake out new authority over Internet network interconnections.[78] Previously, the Commission cautiously approached regulating private marketplace agreements governing the exchange of Internet traffic by broadband ISPs. This is likely due to the fact that the Commission's claimed basis of authority and precedents involve interconnection practices by telephone networks – which involve technologies and market conditions significantly different than those applicable to broadband networks. But given the *Open Internet Order*'s rejection of the idea that regulatory intervention should be tied to findings of market power or consumer harm,

---

74 Michelle Di Ionno & Michael Mandel, "Investment Heroes 2016: Fighting Short-termism," Progressive Policy Institute, at 6 (October 2016), available at: http://www.progressivepolicy.org/wp-content/uploads/2016/10/InvestHeroes_2016.pdf.

75 *Nineteenth Wireless Competition Report*, at 20, ¶ 24 (internal cite omitted).

76 Michael J. Horney, "Broadband Investment Slowed by $5.6 Billion Since Open Internet Order," FSF Blog (May 5, 2017), available at: http://freestatefoundation.blogspot.com/2017/05/broadband-investment-slowed-by-56.html.

77 Hal Singer, "Does The Tumble In Broadband Investment Spell Doom For the FCC's Open Internet Order," *Forbes.com* (August 25, 2015), available at: http://www.forbes.com/sites/halsinger/2015/08/25/does-the-tumble-in-broadband-investment-spell-doom-for-the-fccs-open-internet-order/#5f3082ee2627.

78 *Open Internet Order*, at 86-96, ¶¶ 194-206.

the Commission handed itself power to regulate private interconnection without principled criteria to limit agency discretion. Indeed, the Commission made settlement-free interconnection a condition for merger approval in its *Charter/Time Warner Cable Order* (2016). That order devoted three pages to minutely detailing the required terms of Charter Communications' post-merger interconnection agreements.[79]

The Federal Trade Commission's (FTC) jurisdiction over privacy for broadband consumers was effectively removed on account of the FCC's Title II classification of broadband services.[80] Then the FCC's *Broadband Privacy Order* (2016) doubled down on its removal of FTC jurisdiction over Internet privacy.[81] Under rules established by the order, broadband service providers – and not online edge providers such as Google or Amazon – were subjected to broad "opt-in" requirements that discourage broadband service providers from offering consumers targeted marketing deals, selling advertisements to personally design consumer experiences, or offering sponsored data as well as free data or zero-rated plans – all of which potentially benefit them.[82] The Commission's new privacy regulations also banned certain "financial inducement practices," such as offering discounts for use of personal information, which could deprive consumers of their choice to enjoy free or inexpensive services.[83] Its new privacy regulations assumed consumers are not competent to decide for themselves whether they would like to make money or personal information the form of exchange for services. And the Commission provided no evidentiary basis for its implicit assumption that consumers

---

79   *Charter/Time Warner Cable Order*, at 214-216.

80   *Open Internet Order*, at 16-17, ¶¶ 220-224; *id.*, at 53-54 ¶¶ 462-467.

81   FCC, Protecting the Privacy of Customers of Broadband and Other Telecommunications Services, WC Docket No. 16-106, Report and Order ("*Broadband Privacy Order*") (released November 2, 2016), available at: https://apps.fcc.gov/edocs_public/attachmatch/FCC-16-148A1.pdf.

82   *Broadband Privacy Order*, at 66-95, ¶¶ 166-233.

83   *Broadband Privacy Order*, at 124-129, ¶¶ 294-303.

in markets characterized by convergence around digital, IP-based services want different sets of basic data privacy protections to apply depending upon whether they are doing business with a broadband service provider or edge provider.

Ultimately, Congress passed a resolution to repeal the Commission's new privacy regulations pursuant to the Congressional Review Act.[84] The resolution was signed by President Trump on April 3, 2017.[85] Chairman Pai announced his intent for the Commission to establish a framework for protecting consumers' online privacy that is consistent with the FTC's framework.[86] Yet he also expressed his view that "the best way to achieve that result would be to return jurisdiction over broadband providers' privacy practices to the FTC, with its decades of experience and expertise in this area."[87]

In May 2017, the FCC adopted a proposed rulemaking to reclassify broadband Internet access services as Title I "information services," eliminate the vague general conduct standard, revise the order's bright-line rules, and restore the FTC's jurisdiction over broadband providers' privacy practices.[88] If adopted, the

---

84   U.S. Congress. Senate. *A joint resolution providing for congressional disapproval under chapter 8 of title 5, United States Code, of the rule submitted by the Federal Communications Commission relating to "Protecting the Privacy of Customers of Broadband and Other Telecommunications Services,"* 115th Cong. 1st sess. S.J.R. 34, available at: https://www.congress.gov/bill/115th-congress/senate-joint-resolution/34.

85   The White House, Office of the Press Secretary, Press Release: "President Donald J. Trump Signs H.J. Res. 69, H.J. Res. 83, H.R. 1228, S.J.Res. 34 into Law" (April 3, 2017), available at: https://www.whitehouse.gov/the-press-office/2017/04/03/president-donald-j-trump-signs-hjres-69-hjres-83-hr-1228-sjres-34-law.

86   FCC Chairman Ajit Pai, "Statement of FCC Chairman Ajit Pai on Congressional Resolution of Disapproval of FCC Broadband Privacy Regulations" (March 28, 2017), available at: http://transition.fcc.gov/Daily_Releases/Daily_Business/2017/db0328/DOC-344116A1.pdf.

87   FCC Chairman Ajit Pai, "Statement of FCC Chairman Ajit Pai on Congressional Resolution."

88   FCC, Restoring Internet Freedom, WC Docket No. 17-108, Notice of Proposed Rulemaking (released May 23, 2017), available at: https://apps.fcc.gov/edocs_public/attachmatch/FCC-17-60A1.pdf.

proposed rulemaking would roll back the regulatory overreaches of the 2015 *Open Internet Order* and the 2016 *Broadband Privacy Order* and restore a light-touch regulatory approach that promotes innovation, investment, and consumer welfare. However, as will be discussed further, restoration of a light-touch regulatory policy by the current FCC will not safeguard sufficiently flexibility and freedom for Internet service providers for the long-term. An act of Congress is necessary to do that.

Aside from the FCC's aggressive imposition of new regulations over broadband – now hopefully to be reversed – other areas of the communications marketplace were targeted for restrictions between 2014 and 2016. The Commission appeared likely to impose new regulation on mobile messaging. In October 2015, the Commission circulated for public comment a petition seeking an FCC declaration that text messaging and other mobile messaging services are "telecommunications services" and thus subject to Title II common carrier regulation.[89]

Although never previously subjected to regulatory classification, if anything, these mobile messaging services fit squarely within the definition of lightly regulated Title I "information services." They unmistakably involve "a capability for generating, acquiring, storing, transforming, processing, retrieving, utilizing, or making available information via telecommunications."[90] Of course, text and other mobile messaging services have flourished in the absence of regulation. Americans sent approximately 1.66 trillion text messages and 277.9 billion multimedia messages were sent in 2016.[91] And no evidence of market power or consumer

---

89　FCC, Petition of Twilio Inc. For An Expedited Declaratory Ruling Clarifying the Regulatory Status of Mobile Messaging Services, Public Notice, WT Docket No. 08-7 (October 13, 2015), available at: https://apps.fcc.gov/edocs_public/attachmatch/DA-15-1169A1.pdf.

90　47 U.S.C. § 153(20).

91　CTIA, "Annual Wireless Industry Survey" ("CTIA's Wireless Industry Summary Report, Year-End 2016 Results")(May 2017), available at: https://www.ctia.org/industry-data/ctia-annual-wireless-industry-survey.

harm has been shown that would justify new restrictions for messaging services. The wireless messaging services market is competitive and consumers have choices not only among messaging services provided by wireless carriers but among wirelessly-accessible IP-based alternatives, including instant messaging, social media, and email. Common carrier regulation would make it more difficult for wireless providers to implement new business models designed to restrict spam and other unwanted and sometimes harmful messages to consumers.

Meanwhile, the FCC aggressively pushed for new regulations controlling how video set-top box devices and video applications function. Section 629 of the Telecommunications Act of 1996, adopted in the era of VCRs and analog cable systems, provides the Commission with limited authority to ensure commercial retail availability of cable "converter boxes" and other equipment used by consumers to access MVPD services.[92] Yet without bothering to undertake any cost-benefit analysis, the Commission launched a plan to force cable, satellite, and other video subscription services to re-engineer ways their networks deliver video and alter their business contractual arrangements. Its "unlock the box" or "AllVid 2.0" proposal would have required video service providers to deliver copyrighted video content and related service information to third-party device makers who never negotiated licenses with the video content copyright owners. The proposed rules would have allowed third-party device makers to take these information flows and repackage them with add-ons or perhaps advertisements for viewing on their retail devices.

Subsequently, the FCC hastily revised its proposal – call it AllVid 2.1? – which it never made publicly available. Chairman Tom Wheeler insisted the revised proposal somehow would ensure that copyright protections would be honored by third parties who never signed agreements with copyright

---

92   47 U.S.C. § 549.

owners.[93] More circumspectly, the Copyright Office observed in a letter to a bipartisan group of members in Congress that the proposal provided no enforcement mechanism to back up the Commission's claims.[94] Indeed, the Copyright Office declared that the proposed video device rules "appear to inappropriately restrict copyright owners' exclusive rights to authorize parties of their own choosing to publicly perform, display, reproduce, and distribute their works according to agreed conditions, and to see remuneration for additional uses of their works."[95]

The risk to protecting intellectual property rights in video programming is one of many reasons why the FCC's device regulation proposal was seriously problematic. There is also serious likelihood that the technical mandates like the ones proposed by the Commission would artificially prolong the life of video set-top boxes and delay market transition to an apps-centric model. The emergence of MVPD competition and rivalrous OTT services along with breakthrough digital content delivery technologies, including streaming media devices and video apps, has completely upended Section 629's purpose. Consumer choice among independent alternatives – including online video services and digital streaming devices – and the growing market trend toward video apps for content viewing made the Commission's heavy-handed proposals particularly unjustifiable. Having previously voiced many such concerns about new FCC regulations for video devices and apps, shortly into his tenure Chairman Ajit Pai removed the revised regulatory proposal from circulation among the Commissioners.

---

93   *See, e.g.,* John Eggerton, "FCC's Wheeler Still Plans September Set-Top Vote," *Broadcasting & Cable* (September 16, 2016), available at: http://www.broadcastingcable.com/news/washington/fccs-wheeler-still-plans-september-set-top-vote/159687.

94   Letter to United States Representatives Blackburn, Butterfield, Collins and Deutch from Maria A. Pallante, The Register of Copyrights of the United States of America ("*Copyright Office Letter*") (August 3, 2016), available at: http://blackburn.house.gov/uploadedfiles/co_set-top_letter.pdf.

95   *Copyright Office Letter.*

These same competitive and technological developments in the market for video services rendered the FCC's December 2014 proposal to redefine "multichannel video programming distributor" services to include OTT or OVD services particularly misguided.[96] By redefining MVPD services to encompass online video services like Hulu, Amazon Prime, or Netflix, the Commission would have brought them under the legacy regulatory framework – or at least a modified version of it. As previously described, the legacy framework was designed to address perceived monopoly-like conditions in the 1990's cable video market. Online video technologies and services were developed in a competitive free market environment, unrestrained by sector-specific mandates such as program access or program carriage requirements. Those online video developments support a case for dismantling the legacy framework, not expanding it. Indeed, jettisoning legacy regulations would allow traditional MVPDs to better respond to OTT competition and meet changing consumer demands with more flexible and more attractively priced service options. Fortunately, the MVPD redefinition proposal has not been adopted.

The FCC also ramped up its decade-long pursuit of new rate regulations governing so-called "special access" or newly-renamed "business data" (BDS) services. These types of services involve private or dedicated facilities that cater to specific high-volume users, typically business enterprises, with quality performance guarantees. BDS services are almost always business-to-business specialized arrangements involving sophisticated parties, not ordinary residential consumers.

Historically, special access lines were provided almost exclusively by incumbent local exchange carriers using copper

---

96   FCC, Promoting Innovation and Competition in the Provision of Multichannel Video Programming Distribution Services, MB Docket No. 14-261, Notice of Proposed Rulemaking (released December 19, 2014), available at: https://apps. fcc.gov/edocs_public/attachmatch/FCC-14-210A1.pdf.

wire-based transmission technologies. Beginning in 1999 the Commission relaxed some of its rate controls for those services in light of emerging competition. Telephone company providers subsequently rolled out advanced technologies, including packet-switched fiber-optic broadband. Market entry and expansion by cable operators and 4G LTE capabilities also provide business enterprises additional choices. Despite these competitive developments and relying on dubious rationales, the Commission zealously pursued new rate controls on BDS services. New regulations would benefit only a narrow segment of competitors who would rather lease their rivals' facilities at government-set rates and also undermine incentives for future investment in BDS facilities. Thankfully, the Commission's proposal to impose new rate controls on BDS services was also withdrawn by Chairman Ajit Pai. Under Chairman Pai's leadership, the Commission has instead adopted reforms that will reduce price controls, including by permitting pricing flexibility for BDS incumbents in geographic areas that satisfy a competitive market test.[97]

The FCC's actions regarding universal service have intensified concerns about the Commission's unwillingness to ensure fiscal responsibility. In adopting the FCC's *USF Reform Order* (2011),[98] members of the Commission expressed hope for constraining the growth of the fund. Yet USF subsidy spending continues to grow – and threatens to balloon. In 2016, total authorized USF spending totaled $8.75 billion.[99] Of that amount, High-Cost Program distributions to eligible carriers serving less densely

---

97   FCC, Business Data Services in an Internet Protocol Environment, WC Docket No. 16-143, *et al*, Report and Order (released April 28, 2017), available at: http://transition.fcc.gov/Daily_Releases/Daily_Business/2017/db0428/FCC-17-43A1.pdf.

98   FCC, Connect America Fund, WC Docket No. 10-90, *et al*, Report and Order and Further Notice of Proposed Rulemaking ("*Universal Service Reform Order*") (released November 18, 2011), available at: http://transition.fcc.gov/Daily_Releases/Daily_Business/2012/db0206/FCC-11-161A1.pdf.

99   2015 USAC Report, at 40.

populated areas totaled nearly $4.6 billion,[100] an all-time high for the program. Moreover, the Commission's *E-Rate Modernization Order* (2014) directed $1.5 billion in additional funding for the USF's E-Rate program to schools and libraries.[101] This increased phone service fees $1.90 per year per phone line in order to pay for the projected $3.9 billion annual price tag for E-Rate.[102]

Rising USF spending has made broadband Internet access services an inviting target for fund contributions. By reclassifying broadband service as a Title II telecommunications service, the Commission paved the way for broadband services – that is, broadband consumers – to be subject to USF surcharges just like any other Title II service. The *Open Internet Order* noted what it considered to be administrative benefits in subjecting broadband services to USF surcharges.[103] And the order reiterated that the Commission referred the question of modifying the universal service contribution methodology to the Federal-State Joint Board on Universal Service for a recommendation.[104] The implication of the order was that the Joint Board would be tasked to make a recommendation on subjecting broadband consumers to USF surcharges.

In addition, the FCC's misreading and misuse of Section 706 of the Telecommunications Act as a stealthy source of extraordinary open-ended regulatory power continued. The Commission pinned its *Municipal Broadband Preemption Order* (2015) to its far-fetched reading of that provision.[105] As a matter of constitutional

---

100   2015 USAC Report, at 40.

101   FCC, Modernizing the E-rate Program for Schools and Libraries, WC Docket No. 13-184; Connect America Fund, WC Docket No. 10-90, Second Report and Order and Order on Reconsideration (*"E-Rate Modernization Order"*) (released December 19, 2014).

102   *E-Rate Modernization Order*, at 47-50, ¶¶ 114-118.

103   *Open Internet Order*, at 235-236, ¶ 489.

104   *Open Internet Order*, at 236, ¶ 489 note 1471.

105   FCC, City of Wilson, North Carolina Petition for Preemption of North Carolina General Statute Sections 160A-340 et seq., WC Docket No. 14-115; The Electric Power Board of Chattanooga, Tennessee Petition for Preemption of a Portion

federalism, the FCC's attempt to supersede state law restrictions on local government ownership of broadband networks was one of the most significant and problematic preemptive actions ever attempted in the Commission's history. Among its effects, the *Municipal Broadband Preemption Order* would have given local governments a right to offer broadband services outside their own territorial boundaries. This would have conferred on local governments powers never bestowed by the states that created them, effectively rearranging state governmental decision-making. In August 2016, the U.S. Court of Appeals for the Sixth Circuit held that Section 706 did not authorize the preemption attempted by the FCC.[106] On this occasion, at least, the Commission was held accountable to fundamental constitutional constraints regarding federalism.

Deregulatory Defaults: FCC Precedents to Build On

To its credit, the FCC finally acknowledged in its *Non-Dominance Order* (2016) that "incumbent LECs no longer presumptively exert market power in their provision of [interstate switched access] services, so dominant carrier treatment under certain of our rules is no longer warranted."[107] It is by means of these services that interexchange carriers gain access to local telephone exchanges in order to complete long distance calls. However, the Commission's

---

of Tennessee Code Annotated Section 7-52-601, WC Docket No. 14-116, Memorandum Opinion and Order (*"Municipal Broadband Preemption Order"*) (released March 12, 2015), available at: https://apps.fcc.gov/edocs_public/attachmatch/FCC-15-25A1.pdf.

106   *Tennessee v. FCC*, 832 F.3d 597 (6th Cir. 2016).

107   FCC, Technology Transitions, GN Docket No. 13-5; USTelecom Petition for Declaratory Ruling That Incumbent Local Exchange Carriers Are Non-Dominant in the Provision of Switched Access Services, WC Docket No. 13-3; Policies and Rules Governing Retirement of Copper Loops by Incumbent Local Exchange Carriers; RM-11358; Declaratory Ruling, Second Report and Order, and Order on Reconsideration (*"Non-Dominance Order"*), at 4, ¶ 8, (released July 15, 2016), available at: https://apps.fcc.gov/edocs_public/attachmatch/FCC-16-90A1.pdf.

grant of modest relief from dominant carrier restrictions and increased allowance for market competition for switched access services was long past due. Even at the time USTelecom filed its petition for relief in December 2012, there was no evidence of market power for those services but ample evidence of competition from rapidly growing VoIP connections and wireless alternatives to traditional switched access lines. The Commission's unjustifiable delay in granting the requested relief meant subjecting ILECs to continued regulatory compliance costs that diverted financial resources from investment in broadband network deployments and upgrades. And such needlessly prolonged legacy regulation offered consumers no demonstrable benefits.

The *Effective Competition Order* (2015) is perhaps the one rally point for genuine market-based reform adopted by the FCC since the chapters in this book were written.[108] In its *Effective Competition Order*, the Commission readjusted its outdated cable regulations upon concluding that nearly every local market nationwide enjoys effective competition for video services. Among other things, the Commission cited DBS's nationwide market share of over one-third of all video subscribers. Based on actual evidence of market share and competing service options, the Commission established a rebuttable presumption that local cable markets are subject to effective competition.[109] This presumption provided a streamlined and low-cost pathway for granting relief to cable operators from local rate regulations of basic tier cable services and leased consumer premises equipment.[110]

---

108 FCC, Amendment to the Commission's Rules Concerning Effective Competition; Implementation of Section 111 of the STELA Reauthorization Act, MB Docket No. 15-53, Report and Order ("*Effective Competition Order*") (released June 3, 2015), available at: http://transition.fcc.gov/Daily_Releases/Daily_Business/2015/db0610/FCC-15-62A1.pdf; affirmed by *National Association of Telecommunications Officers Association v. FCC*, No. 15-1295 (D.C. Cir. July 7, 2017).

109 *Effective Competition Order*, at 4-6, ¶ 6.

110 *Effective Competition Order*, at 14-19, ¶¶ 17-26.

A modernized Communications Act for the Digital Age should build on the type of presumptive deregulatory approach exemplified by the *Effective Competition Order*. A presumption of marketplace freedom should become an organizing principle for a modernized Communications Act, as discussed in further detail in Chapter 6. Adoption of a presumption of marketplace freedom places the burden of demonstrating the need for regulation on the FCC or on petitioners advocating regulation. A presumptive approach favoring market freedom should be applied to digital communications services across the board as part of a comprehensive, clean-slate reform of federal communications law. Thus, under a statute such as we advocate here, the FCC never could have adopted the *Open Internet Order*, or if so, such order would not survive judicial review. Concomitantly, the Commission never could have adopted privacy regulations that were so discordant from the realities of the competitive marketplace and consumer welfare. And, considering the competitiveness of the video marketplace, it would have been a non-starter for the Commission even to propose new regulatory mandates for TV set-top boxes or video apps.

Congress Must Prohibit Public Utility-Style Regulation of Broadband Services

It is imperative that public utility-style regulation of broadband Internet access services not only be eliminated, but prohibited from being re-adopted by a differently configured FCC in the future. And it is also important that the FTC regain its role as the general enforcer of online privacy practices. Even if the FCC repeals its misguided 2015 *Open Internet Order* – which it should – the future of digital communications services remains vulnerable to renewed regulatory overreach by the Commission. Congress should adopt legislation to provide long-lasting resolutions to both broadband and online privacy policy issues triggered by the order.

Congress should adopt legislation to restore and sustain Internet freedom, innovation, and investment, while also providing a carefully targeted means for the FCC to address specific instances of consumer harm and anticompetitive conduct by broadband Internet service providers. Congress should establish the FTC as the agency responsible for consumer privacy across all online platforms and services. Certainly, Congress could choose to pursue these objectives through separate targeted legislative measures. For example, it may make sense for Congress to adopt legislation that would end, at least for the foreseeable future, the long-running contentious battle over "net neutrality." But it is important such legislation be grounded in the principles addressed in this book.

Even if narrowly-focused legislation is adopted in the near-term to address, say, the ongoing net neutrality controversy, a more comprehensive, clean-slate reform of the nation's communications laws offers a superior vehicle by which Congress can provide long-term resolutions to fundamental policy issues regarding broadband, online privacy, video, and other communications services. Ultimately, a modernized Communications Act is essential to providing a durable free market, rule of law-oriented framework governing communications services for the Digital Age.

Conclusion

We hope that policymakers, academics, analysts, and interested citizens alike will find the information and insights offered in this book useful for understanding the pressing need for Congress to undertake adoption of a modernized Communications Act. The glaring mismatch between the outdated assumptions of legacy regulation embodied in the current statute – or at least not dispelled or precluded by the current Act – and the realities of today's dynamic communications market finally should be alleviated by the type of clean-slate reform proposed in this book.

The threat posed to broadband's future – and to the Internet's future – by the FCC's historical pro-regulatory institutional bias must be addressed by reform. Congress should now seize the opportunity to establish a framework that itself curtails and controls what has been called the "bureaucratic imperative." Legacy regulatory paradigms that no longer make sense and that are now counterproductive and costly must give way to a new legislative framework embodying deregulatory presumptions and antitrust-like jurisprudence rooted in an analysis regarding the existence or non-existence of a market failure. The new statute should be grounded in consumer welfare concerns, not competitive welfare protectionism. The chapters contained in this book comprise a roadmap – or a "how to" manual if you prefer – for moving ahead with a congressional overhaul of federal communications law – a #CommActUpdate – in accordance with free market and rule of law principles.

Congress should act in the near-term to adopt a Digital Age Communications Act. And if it simply cannot accomplish a comprehensive overhaul in the near-term, in the meantime it should pursue a legislative solution to the long-running disputes concerning net neutrality practices that incorporate the free market-oriented principles put forward in this book.

Chapter 1
Summary of the First White Paper

"Modernizing the Communications Act"

"Modernizing the Communications Act" was the first in the series of white papers issued by the U.S. House Committee on Energy and Commerce as part of its plan to examine and update the law to better fit the communications marketplace of today. The paper began with a brief history of communications laws passed by Congress and described basic structure of current federal law. It solicited public comment in response to a series of questions about larger thematic concepts upon which the Communications Act should be updated.

Link: http://energycommerce.house.gov/sites/
republicans.energycommerce.house.gov/files/analysis/
CommActUpdate/20140108WhitePaper.pdf

Response to Questions in the First White Paper

"Modernizing the Communications Act"

by

Randolph J. May, President, The Free State Foundation

and

Members of the Free State Foundation's Board of Academic Advisors:

Richard Epstein, New York University Law School

Justin (Gus) Hurwitz, University of Nebraska College of Law

Daniel Lyons, Boston College Law School

James Speta, Northwestern University Law School [*]

before the

Committee on Energy and Commerce

U.S. House of Representatives

January 31, 2014

---

[*] While the signatories to this Response are in general agreement with the views expressed in these comments, their participation as signatories should not necessarily be taken as agreement on every aspect of the submission. The views expressed should not be attributed to the institutions with which the signatories are identified. This Response is reproduced here in its original form, save for only minor non-substantive edits.

I. Introduction

At the outset, we want to commend the Committee for initiating this process to review and update the current Communications Act. In our opinion, the review is timely because the Communications Act does need updating. And we commend the deliberative and open nature of the review process as it begins.

At the time the review process was announced, Chairman Walden stated: "When the Communications Act was updated almost 18 years ago, no one could have dreamed of the many innovations and advancements that make the Internet what it is today. Written during the Great Depression and last updated when 56 kilobits per second via dial-up modem was state of the art, the Communications Act is now painfully out of date."[111] In the first White Paper, "Modernizing the Communications Act,"[112] the Committee explains the initiation of the updating process this way: "Changes in technology and the rate at which they are occurring warrant an examination of whether, and how, communications law can be rationalized to address the 21st century communications landscape." Regardless of precisely how this proposition is framed, and we do not intend to belabor the matter, the essential point is this: Since the Communications Act was last revised in any meaningful way in 1996, the communications and information services marketplace environment, driven in significant part by technological changes, has changed dramatically. Thus, in our view, the review and updating process not only is timely but necessary.[113]

---

111   "Upton and Walden Announce Plans to Update the Communications Act," News Release, Committee on Energy and Commerce, December 3, 2013.

112   White Paper No. 1, "Modernizing the Communications Act," January 8, 2014.

113   While we certainly hope that the review and updating process proceeds apace in light of the dramatic technological and marketplace changes that already have occurred, we understand that it may be several years before the end of the road is reached. In the interim, Congress should not necessarily refrain from adopting certain targeted revisions that may improve communications policy

The Committee has adopted a wise approach by initially seeking responses to questions that, as the White Paper puts it, "address thematic concepts" for updating the Communications Act. It is certainly preferable to begin the review and public comment process by eliciting responses at this higher thematic level, and then, as the process progressively moves forward, the Committee will be in a better position to seek responses to questions at less thematic, more specific, levels of detail.

The questions asked in the first White Paper are the proper ones. Given the nature of the current statute, and the direction that changes in the statute likely should take, it is not surprising that, at least from our perspective, the questions would elicit responses with considerable overlap and duplication if each is answered separately and in serial fashion. Therefore, we prefer, in order to avoid undue repetition and duplication, to provide a narrative that, in essence, takes the second question: "What should a modern Communications Act look like?" as the primary focal point of our response. In answering this broad framing question posed by the White Paper, we necessarily will address the other questions relating to the structure of the Act, the need for flexibility and technological neutrality, the distinction between information and telecommunications services, and so forth.

---

and which are consistent with the overall market-oriented reform direction that communications policy should take. Examples of such targeted measures that might be appropriate include two bills introduced last year by Rep. Bob Latta, Communications and Technology Subcommittee Vice-Chair: H.R. 2649, the "FCC 'ABCs' Act of 2013," which would revise the forbearance provision in Section 10 of the Communications Act to require clear and convincing evidence that the forbearance requirements are not met before denying relief, and H.R. 3196, the Consumer Choice in Video Devices Act, which would revise Section 629 of the Act to require elimination of the cable television set-top box integration ban imposed by the FCC.

II. The Major Guiding Principles for Reforming the Communications Act

In this section, before providing a more expansive narrative statement responsive to the Committee's questions, we want to set forth in summary fashion what we believe should be the guiding foundational principles of the reform effort. These principles will guide the narrative statement that follows this Section II.

A. **In updating the Communications Act, a clean slate approach is needed to adopt a "replacement" regime – a new Digital Age Communications Act, if you will[114] – because the new act should be much different in concept and structure than the existing one.**

B. **The existing "silo" regime, which results in the regulation of entities providing comparable services in a disparate manner, should be eliminated.**

C. **Generally, the broad delegation of indeterminate authority to the FCC to regulate "in the public interest" should be eliminated.**

---

114   In 2005, at the Progress & Freedom Foundation, Randolph May and James Speta, working with a group of scholars with diverse views and political leanings, led an effort to develop a regulatory framework for what was called a "Digital Age Communications Act" or "DACA." The framework ideas presented here certainly owe much to the ideas developed then, because we think that they remain proper guiding principles at this time. Indeed, we still favor "Digital Age Communications Act" as the name for the new act. Once again, we acknowledge the debt owed to the DACA Working Group. See Randolph J. May and James B. Speta, "Digital Age Communications Act," Proposal of the Regulatory framework Working Group, Progress & Freedom Foundation, June 2005. See also, Randolph J. May, *Why Stovepipe Regulation No Longer Works: An Essay on the Need for a New Market-Oriented Communications Policy*, 58 Fed. Comm. L. J. 103, 106 (2006)(referring to the need for a new regulatory framework that reflects today's digital age competitive marketplace realities, "what one might call a new Digital Age Communications Act").

D. A competition standard should be adopted that is generally applicable to all entities providing electronic communications subject to the Commission's jurisdiction, replacing the existing "silo" regime and "public interest" standard.

E. The FCC's authority to adopt broad anticipatory rules on an *ex ante* basis should be substantially circumscribed, and agency rules should be sunset after a fixed number of years absent a strong showing at the sunset date that they should be continued; the Commission should be required to rely more heavily than is presently the case on adjudicating individual complaints alleging specific abuses of market power and consumer harm.

F. To a significant extent, the FCC's structure as a matter of form in an institutional sense will be dictated by the structure of the new act and the fundamental decisions made regarding the agency's role. The new act should require that the agency adhere to certain process reforms. With respect to jurisdiction, certain matters currently under the FCC's jurisdiction should be transferred to the FTC. Finally, the authority of the states to engage in economic regulation of service providers should be circumscribed in the new act.

In drafting a new act, one guided by these foundational principles, the concept of "simplicity" should remain an important goal. In the Fourteenth Century, William of Ockham wrote: "What can be explained on fewer principles is explained needlessly by more." This theorem became known as Ockham's Razor. In drafting a new act, the Razor should be kept close at hand.

III. "What Should a Modern Communications Act Look Like?"

As stated above, within the context of answering this broad framing question we believe we will answer below all of the questions posed by the Committee. But because we understand that this is just the beginning of the process, a process that certainly will focus more narrowly on specific subject matters and issues as it continues, our response, by design, is broadly thematic in keeping with the nature of the initial set of questions.

A. A Clean Slate for Adopting a Replacement Regime

Perhaps the most fundamental question the Committee will face is whether Congress should approach the updating process by, for the most part, starting with a "clean slate" to devise a replacement regime, or whether, on the other hand, it could achieve what needs to be accomplished in an update by employing more targeted revisions to the current statute. There may well be some who suggest that rather than replacing the current act with a new one embodying a very different model, a principal drafting objective should be to amend the statute as little as possible. We do not discount the possibility that the existing Communications Act could be improved, perhaps even substantially, by amendments to the current statute.[115] But the option of "tinkering around the edges" in order in an effort to minimize the changes to the current statute should be rejected in favor of adopting a replacement statue.

There are two primary reasons for this. First, as explained below, the conceptual changes in communications law and policy

---

115   And as stated in footnote 113, we do not suggest that, in the interim, pending adoption of a comprehensive rewrite, the current act should not be revised in limited targeted ways that are consistent with the market-oriented, less regulatory direction the new act ultimately should take.

that are warranted, indeed required, by the dramatic technological and marketplace changes described in the Committee's White Paper, are major. The governing concepts and philosophical principles embodied in the new act that we envision are very different from the governing concepts and philosophical principles embodied in the current statute. After all, in many important respects, the current statute remains intact as adopted in 1934, and the 1934 act itself closely resembled, in significant respects, the Interstate Commerce Act of 1887. The ICA's very purpose was to tame what were considered to be static carriers exercising monopolistic power, not to oversee a technologically dynamic marketplace. This being so, the "clean slate" approach simply makes more sense.

Second, and relatedly, the clean slate approach is more susceptible to achieving the goal of simplicity. A clean slate approach adopting a replacement regime is much more likely to result in a governing statute that is shorter, better organized, more intelligible, with fewer unintended conflicts, than one that takes the current act as its starting point.[116]

B. The Silos Should Be Eliminated

As the Committee's first question states: "The current Communications Act is structured around particular services." No one really disputes this assertion, and there is fairly widespread agreement that the current act's structure, with its various regulatory "silos" or "stovepipes" is increasingly problematic in the digital age. Under the existing statute, disparate regulations often apply

---

116   It is worth noting here that we understand that there will be attention paid to the need for transitional periods and transitional mechanisms to get from the existing statute to a new one. These transitional issues, which may involve phase-outs of certain obligations and duties over a period of time, will present some difficult and important questions. Nevertheless, at bottom, they are still transitional issues. It would be a mistake to sacrifice the benefits of long-term improvements in the law because of a fear of short-term disruption.

to services that, from the consumer's perspective, compete against each other in the marketplace. Thus, for example, "telecommunications" providers are regulated differently from "information services" providers. "Cable" television operators are regulated differently from broadcasters and "satellite" television operators. Wireless service providers are subject to their own set of regulatory requirements, even though the services they offer increasingly compete with all of the others.

Without belaboring the point, whatever the merits of the "silo" structure in an earlier age, it no longer makes sense. The various silos, in essence, primarily are based on "techno-functional" constructs[117] that do not comport with the realities of digital age technologies and services. Even a casual examination of the definitions of "telecommunications," "information services," "cable service," "mobile service," and so forth shows that these definitions, with the attendant regulatory classification impacts, mostly are tied to descriptions of certain technological capabilities or functional characteristics of the services.

The old saw that a "byte is a byte is a byte" is now a digital world truism, at least in the sense that is relevant here. And it is this technological reality that has rendered the current silo regime obsolete as a policy paradigm as the transition to IP networks has rapidly accelerated. As Christopher Yoo put the matter as early as 2003: "Gone are the days in which each communications

---

117 Randolph J. May, *Calling for a Regulatory Overhaul, Bit by Bit*, CNET News, Oct. 19, 2004, http://news.com./Calling+for+a+regulatory+over-haul%2C+bit+by+bit/ 2010-1028_3-5415778.html. ("The policy framework embodied in our existing communications laws is often called 'stovepipe' regulation. This is because there are distinct technology-based and functionally driven regulations that apply in a disparate fashion, depending on whether different services are classified as telecommunications, information services, cable, satellite or broadcast. Imagine each distinct service classification as a vertical stovepipe…. [O]n the regulation of VoIP services, I called the distinctions underpinning stovepipe regulation metaphysical in the sense that the existing definitions rest upon transcendent and highly abstruse techno-functional constructs.")

technology could be regarded as occupying a separate regulatory silo. The impending shift of all networks to packet-switched technologies promises to complete the collapse of any remaining attempt to base regulation on differences in the means of transmission."[118] Shortly thereafter, in 2004, the FCC itself recognized the impact and implications of the IP transition. In opening its (yet to be completed) *IP-Enabled Services* proceeding, the Commission explained that the greater bandwidth of digital broadband services encourages the introduction of services "which may integrate voice, video, and data capabilities while maintaining high quality of services."[119] Then, in a prediction that certainly has proven to be true, the FCC added: "[I]t may become increasingly difficult, if not impossible, to distinguish 'voice' service from 'data' service, and users may increasingly rely on integrated services using broadband facilities delivered using IP rather than the traditional PSTN (Public Switched Telephone Network)."[120]

The Commission's 2004 statement predicting the increasing difficulty in distinguishing "voice" from "data" services, not at all surprisingly, has been proven true, and this surely is a contributing reason as to why the Commission, some nine years later, has yet to take further action in the *IP-Enabled Services* proceeding definitively to classify interconnected VoIP services as "telecommunications services" or "information services." We are entirely sympathetic to the difficulty faced by the Commission and understand its reluctance to conclude that these IP-based voice services are "telecommunications" with all the attendant regulatory consequences that may flow from such a determination. Nevertheless, it is the consequences of this non-action – at best, continued regulatory uncertainty, or, at worst, the application of

---

118   Christopher S. Yoo, *New Models of Regulation and Interagency Governance*, 2003 Mich. St. L. Rev. 701, 714 (citation omitted).

119   IP-Enabled Services, Notice of Proposed Rulemaking, 19 F.C.C.R. 4863, para. 16 (2004).

120   Id.

disparate regulations to services that, from the consumer's perspective, compete against each other in the marketplace – that demonstrate the need to dismantle the silo regime, including the distinction between "telecommunications service" on the one hand and "information services" on the other.[121]

The transition from narrowband to broadband and from analog to digital has rendered the silo regime statutory structure obsolete. In the current environment, the communications marketplace has become increasingly competitive – but the competition primarily takes place across multiple digital broadband platforms employing various technologies, and sometimes a mix of technologies. These various communications platforms should not be subject to disparate regulations simply because they are consigned to one silo or another. The current approach creates incentives for companies to invest capital in efforts to gain advantages through regulatory arbitrage, rather than investing in ways to deliver better services to consumers more efficiently.

C. The Public Interest Standard Should Be Largely Eliminated

Aside from the silo structure, there is another paradigm in the Communications Act that, for the most part,[122] should be jettisoned in a new Digital Age Communications Act. This is

---

121 This is not to say that the distinction between "telecommunications services" and "information services," which dates back to the *Computer II* regime adopted by the FCC in 1980, did not, for a number of years, play a valuable role in allowing the newly emerging information services to continue to develop free from Title II common carrier regulation. The point is that now, in a new act, the Title II silo itself should be eliminated.

122 We say "for the most part" because there may be limited instances in which retention of the public interest standard might be appropriate. It is not the proper place here, at the beginning of the review process, to attempt to delineate those places, if any. The main point now is the public interest standard is ubiquitous throughout the Communications Act, and this certainly should not be the case in the replacement regime.

the ubiquitous "public interest" standard, which "still pervades the current regulatory regime."[123] There are nearly 100 different provisions in the Communications Act which delegate authority to the FCC to regulate in the "public interest, convenience, and necessity" (or some very close variant thereof).[124] There is an argument that the public interest standard, which originated in Progressive/New Deal era theories of public administration based on notions of an agency's presumed impartial, nonpolitical expertise, is so indeterminate in meaning that it constitutes an unconstitutional delegation of congressional authority.[125] Among long-time FCC-watchers, there is an old saw that the "public interest" is whatever three of the five commissioners say it is on any given day. While the Supreme Court has rejected the constitutional challenge to the public interest standard on nondelegation doctrine grounds,[126] this does not mean that this Progressive/New Deal era standard should remain the FCC's governing lodestar for regulation in today's radically changed environment. It should not. Simply put, the public interest standard is so vague that it necessarily confers too much unbridled discretion on the agency without sufficient direction from Congress.

---

123   See Randolph J. May, A Modest Plea for FCC Modesty, 60 Admin. Law Rev. 895, 897 (2008) ("The public interest standard that was the keystone of the Radio Act of 1927 and its successor, the Communications Act of 1934, still pervades the regulatory regime.")

124   See Randolph J. May, *The Public Interest Standard: Is It Too Indeterminate to Be Constitutional?*, 53 Fed. Comm. L.J. 427, at 456–67 (2001) (listing provisions in the Communications Act that pertain to the public interest standard). In 1999, constitutional law scholar Gary Lawson called the public interest standard "[e]asy kill number 1" on nondelegation doctrine grounds because the licensing provisions of the Communications Act grant "nearly absolute discretion...." Gary Lawson, *Delegation and the Constitution*, Reg., Spring 1999, at 23, 29, *available at* http://www.cato.org/pubs/regulation/regv22n2/delegation.pdf.

125   For a full discussion of this argument with citation to many authorities, see Randolph J. May, *The Public Interest Standard: Is It Too Indeterminate to Be Constitutional?*, 53 Fed. Comm. L.J. 427 (2001).

126   See FCC v. Pottsville Broad. Co., 309 U.S. 134, 138 (1940); National Broadcasting Co. v. United States, 319 U.S. 190, 219-226 (1943).

D. A Marketplace Competition Standard Should Replace the Silos and Public Interest Standard

If the silo regime should be disassembled and the public interest standard largely jettisoned, then what should be at the core of the replacement regime as the governing lodestar? The answer is a competition-based standard that directs the FCC generally to undertake an antitrust-like economic analysis when it engages in regulatory activity that is subject to its jurisdiction.[127] We are not suggesting that a new statute direct the FCC, in an overtly strict sense, to incorporate and apply current antitrust jurisprudence or precedents. But given the development of competition in most communications market segments, and the technological dynamism that characterizes these markets, the Commission generally should be required to find a market fail-ure – that is, the existence of market power – and, concomitantly, demonstrable consumer harm resulting from the exercise of market power, before imposing regulations on entities subject to

---

127 In our view, the Commission generally should retain jurisdiction over electronic communications networks and services that, to a significant extent, mirrors the extent of the agency's jurisdiction under the current Communications Act. Thus, while the language would be updated to reflect modern usage, we envision that the scope of the Commission's *jurisdiction* would not be materially narrowed from the scope of the agency's *jurisdiction* in present Section 2(a) of the Act. We hasten to add, though, as explained in the text, that the *exercise* of whatever jurisdiction the Commission is granted should be substantially constrained by the new competition-based standard that ties the *exercise* of the Commission's authority to findings of market failure and consumer harm. In other words, it is important to distinguish between the conferral of jurisdiction and the constraints placed on the exercise of such jurisdiction. Finally, there are delegations of authority in the current act, such as the enforcement of privacy (CPNI) rules for telephone and cable companies, that probably should be transferred to the Federal Trade Commission so that various entities providing comparable services in the Internet ecosystem would be subject to the same type of regulations. And, enforcing a uniform set of privacy rules, for example, is closer to the core competency of the FTC than the FCC.

its jurisdiction.[128] In line with the recommendation of the Digital Age Communications Act Regulatory Framework Working Group, and the technological dynamism that characterizes the communications marketplace, it may be advisable for the new statute to specify that any market failure found by the Commission must be determined to be "non-transitory."[129]

By virtue of adoption of a competition standard grounded in antitrust-like jurisprudential principles, before regulating the FCC would be required, much more than it is today under the existing act, to engage in a rigorous economic analysis that focuses on actual and potential marketplace competition. As part of such analysis, the agency necessarily would need to take into account the impact of the dynamism that characterizes the digital marketplace.

E. Curtailing *Ex Ante* Rulemakings and Relying More Heavily on *Ex Post* Adjudication of Complaints

In a new act, the FCC should be required to favor narrowly-tailored *ex post* remedial orders in addressing practices that are alleged to be anticompetitive or abusive rather than undertaking broad *ex ante* proscriptions developed in generic rulemakings. The agency generally should be required to determine whether service providers subject to individualized complaints possess demonstrable market power that should be constrained in some appropriately targeted way and whether the exercise of such market power has caused consumer harm in a way that should be subject to regulatory sanction. So, rather than the FCC

---

128   There may be some limited areas of regulatory activity subject to the Commission's jurisdiction that should not be tied to the market failure standard, but the purpose here is to suggest the proper general framework, not to identify any specific exceptions. These may be addressed as the updating process continues.

129   See Randolph J. May and James B. Speta, "Digital Age Communications Act," Proposal of the Regulatory Framework Working Group, Progress & Freedom Foundation, June 2005, and note 114 supra.

embarking on generic rulemaking proceedings – as it currently most often does – that frequently end with the adoption of overly broad proscriptions that result from the perceived (to some extent self-induced) imperative to anticipate harms that may never materialize, regulatory prohibitions and sanctions under the new statute generally would be accomplished through the conduct of focused adjudicatory proceedings following the filing of individual complaints containing specific allegations of abuse of market power.

A marketplace competition standard would make it easier for broadband companies to engage in permission-less innovation. Commentators have long acknowledged that competition improves if entrepreneurs can develop ideas and bring new products to market without first needing to seek government approval. Unfortunately, an *ex ante* regulatory regime often inhibits permission-less innovation by suggesting that new products be submitted to the Commission for review. An *ex post*, competition-based standard would clarify that entrepreneurs are free to introduce new ideas and products to the marketplace without prior regulatory approval, provided that the offering doesn't abuse market power in a way that causes consumer harm.

While the Commission should not necessarily be precluded from adopting rules that define, in advance, certain specific acts or practices that constitute abuses of market power because they cause consumer harm, this rulemaking authority should be carefully circumscribed. Thus, the Commission should not have authority to issue proscriptive regulations unless it determines, based on a showing of clear and convincing evidence in the rulemaking proceeding, that marketplace competition is not sufficient to protect consumers from harm. And its findings should be subject to ongoing review, as befits conclusions drawn about the state of affairs in a dynamic marketplace. To enforce the principle, any such rules the Commission issues regarding competition should automatically sunset after an appropriate period of time, say, for example, in five years, unless the Commission affirmatively finds, again based on a showing of clear and convincing evidence, that there is a market failure necessitating continuation of the regulation in order to prevent consumer harm.

F. The Structure and Jurisdiction of the FCC: Process Reform, Transfer of Jurisdiction to the FTC, and Limits on State Economic Regulation

The Committee asks, quite appropriately, about the structure and jurisdiction of the FCC, and how they should be tailored to address the systematic change in communications. To a significant extent, of course, the structure of the agency, in an institutional sense, should be strongly influenced by – or "follow" as in the saying, "form follows function" – the jurisdiction the agency is granted and the structure of the new act in defining the agency's exercise of its regulatory authority. In other words, in a new Digital Age Communications Act without silos, there likely should not be an FCC, institutionally, with separate Mass Media, Wireline, Wireless bureaus, as opposed to, say, a Broadband Bureau. And in an FCC in which a marketplace competition standard generally replaces the public interest standard as the agency's regulatory lodestar, then from an internal agency organization perspective, the role of economic analysis – and the economists responsible for performing such analysis – should be institutionalized in an appropriate organizational manner that furthers the usefulness and effectiveness of such analysis.

Any new act should contain within it some of what, for present purposes, might be called "process reforms." As Free State Foundation President Randolph May has testified before this Committee twice in the last three years, these reforms should include a range of process improvements, ranging from additional analytical requirements for agency rulemakings to transaction process reforms, to the institution of "shot clocks" for completing agency proceedings and requirements for more input by all commissioners in controlling the Commission's docket.[130] These institutional process reforms are important to making the FCC the "model agency for the digital age" that then-FCC Chairman William Kennard envisioned in 1999, when the agency, under his direction, released a report entitled, "A Strategic Plan: A New FCC for the 21st Century." The plan's first four sentences read:

> In five years, we expect U.S communications markets to be characterized predominately by vigorous competition that will greatly reduce the need for direct regulation. The advent of Internet-based and other new technology-driven communications services will continue to erode the traditional regulatory distinctions between different sectors of the communications industry. As a result, over the next five years, the FCC must wisely manage the transition from an industry regulator to a market facilitator. The FCC as we know it today will be very different in both structure and mission.[131]

Unfortunately, since 1999, there have been few, if any, meaningful changes regarding the structure and mission at the agency. The proposals contained in Chairman Walden's "Federal Communications Commission Process Reform Act of 2013" (H.R. 3675) and in Mr. May's testimony before the Committee certainly should be considered in junction with a new act.

With respect to the FCC's jurisdiction, as lines continue to

---

130   See Testimony of Randolph J. May, Hearing on "Improving FCC Process," Subcommittee on Communications and Technology, July 11, 2013; Testimony of Randolph J. May, Hearing on "Reforming FCC Process," Subcommittee on Communications and Technology, June 22, 2011. [Reprinted in Appendix – *eds.*]

131   FCC, "A Strategic Plan: A Model Agency for the 21st Century," August 1999.

blur across the Internet ecosystem among various providers of services that, from the consumer's perspective, are comparable – regardless whether they presently are denominated as facilities-based network service providers, "over-the-top" providers of VoIP services, or content and applications "edge" providers, or whatever – it will be important in drafting a new act to consider treating such services in a holistic way, at least for some purposes that relate more closely to consumer protection than traditional economic regulation. For example, with regard to any regulatory oversight relative to the protection of privacy or data security, even though the FCC presently has some jurisdiction in these areas, for the most part, it would be preferable to consolidate such jurisdiction in the FTC. The types of consumer protection issues most likely to arise with regard to privacy and data security are at the core of the FTC's institutional expertise. If jurisdiction over these type of matters – matters outside of the purview of traditional economic regulation of service providers – is transferred to the FTC, it is much less likely that telecom and cable services providers, on the one hand, and, say, Facebook or Twitter, on the other, would end up subject to disparate regulations in these areas.

Finally, a new act must also address the role of state regulators in the 21st century telecommunications marketplace. The Communications Act of 1934 divided regulatory authority over telecommunications services between the federal government and the states. This distinction was appropriate when regulating twentieth-century telephone networks, which were primarily regional monopolies that distinguished between local and long-distance calls. But today's information service networks generally are national in scope. Neither providers nor consumers can distinguish easily, if at all, and certainly not without significant costs, between interstate and intrastate communications. The law should not do so either. State regulation of economic activity such as rates or conditions of market entry jeopardizes the economics of scale that flow from national networks. Many states recognize this and have reduced their telephone regulations while foreswearing any interest in regulating IP networks. The new act should similarly vest most regulatory authority in the federal government rather than the states.

But this does not mean states should play no role under a new act. States have a significant advantage over their federal counterparts regarding issues where local knowledge may be brought to bear, and sound policymaking should continue to leverage that advantage. For example, state and local authorities should retain primary jurisdiction over siting decisions, because they know best how specific projects will affect a local community. Similarly, state regulators are in a better position to understand the individualized needs of local communities and thus should retain a prominent voice regarding consumer protection issues, though subject to federal oversight to assure that parochial issues do not needlessly jeopardize broader national objectives.

Chapter 2
Summary of the Second White Paper

"Modernizing U.S. Spectrum Policy"

"Modernizing U.S. Spectrum Policy" was the second white paper issued by the House Committee on Energy and Commerce as part of its plan to examine and update the Communications Act. The paper provided a brief overview of the history of spectrum allocation, regulation, and licensing in the U.S. It solicited public comment in response to a series of questions about specific spectrum issues.

Link: http://energycommerce.house.gov/sites/
republicans.energycommerce.house.gov/files/analysis/
CommActUpdate/20140401WhitePaper-Spectrum.pdf

Response to Questions in the Second White Paper

"Modernizing U.S. Spectrum Policy"

by

Randolph J. May, President, The Free State Foundation

and

Members of the Free State Foundation's Board of Academic Advisors:

Michelle P. Connolly, Duke University

Richard A. Epstein, New York University Law School

Justin (Gus) Hurwitz, University of Nebraska College of Law

Daniel Lyons, Boston College Law School

Bruce M. Owen, Stanford University

Richard J. Pierce, Jr., George Washington University School of Law

James B. Speta, Northwestern University Law School[*]

before the

Committee on Energy and Commerce

U.S. House of Representatives

April 25, 2014

---

[*] While the signatories to this Response are in general agreement with the views expressed in these comments, their participation as signatories should not necessarily be taken as agreement on every aspect of the submission. The views expressed should not be attributed to the institutions with which the signatories are identified. This Response is reproduced here in its original form, save for only minor non-substantive edits.

Randolph J. May and Seth L. Cooper

I. Introduction and Summary

At the outset, we want to again commend the Committee for initiating this process to review and update the current Communications Act. As we said in our initial Response, the review is timely because, as a result of technological advances and dramatic marketplace changes, many of which were described in our first Response,[132] the Communications Act does need updating. And, as part of that process, there is no doubt that spectrum policy needs "modernizing" too.

In preparing this Response, our focus is on offering general principles that should guide Congress's consideration in drafting a modernized approach to spectrum policy and the Federal Communications Commission's approach to implementing such a modernized policy. While the Committee's white paper presents a number of questions on specific topics, we believe, at least at this stage, that the generalized framework presented in these comments offers a more useful approach in responding holistically to the questions than would individual responses to specific questions. Indeed, this approach is consistent with a central theme of our comments: spectrum policy is undergoing its own "convergence" of sorts, and a proper policy framework for spectrum should encourage and anticipate this convergence on a uniform, integrated basis.

To briefly summarize our views: The current administrative fiat approach has its roots in the Radio Act of 1912, which was passed in the wake of the Titanic disaster. It vested authority to issue licenses for wireless communications in the Secretary of Commerce as a means of mitigating interference. After radio broadcast stations began operating in the 1920s – and their broadcasts on

---

132 Free State Foundation Response to Questions in the First White Paper, "Modernizing the Communications Act," January 31, 2014, *available at*: http://www.freestatefoundation.org/images/Response_to_Questions_in_the_First_White_Paper_013114.pdf.

the same or nearby frequencies began to interfere with each other – Congress passed the Radio Act of 1927,[133] which created the Federal Radio Commission ("FRC"). The FRC was succeeded, largely without change, by the Federal Communications Commission ("FCC") with the passage of the Communications Act of 1934.[134]

The basic structure of spectrum regulation, which was initially formulated over 100 years ago and has remained largely the same since adoption of the Radio Act of 1927, is based on technological, economic, and legal views that no longer make sense – to the extent they ever did. It is true that in the past two decades there have been some changes, such as the initiation of auctions to award spectrum licenses, which represent positive steps toward a more market-oriented regime. Nevertheless, a key feature of the administrative model that has prevailed in the last century, and which largely continues to prevail today, relies on allocation by the FCC of particular frequency bands for particular pre-specified service uses in accordance with particular pre-specified technical parameters. This process requires the use of extensive, and often pointless and redundant administrative procedures that have long outlived their usefulness. Regrettably, this "command-and-control" administrative regime dramatically fails to promote flexible use of spectrum. As economics scholar Thomas Hazlett, a former FCC Chief Economist and a leading authority on spectrum policy, has explained, "[t]he weakness of the administrative allocation regime is that it regularly resolves potential conflicts among wireless users by suppressing vast amounts of productive activity."[135] Over the past few decades, there have been substantial technological changes, including the advent of digital wireless

---

133 Pub. Law No. 632, ch. 169, 44 Stat.1162 (1927).

134 Pub. Law No. 416, ch. 652, 48 Stat. 1064 (1934).

135 Thomas W. Hazlett, *A Law and Economics Approach to Spectrum Property Rights, A Response to Weiser and Hatfield*, 15 Geo. Mason L. Rev. 975, 977 (2008).

communications, which have been coupled with the adoption of auctions as the preferred method for the initial assignment of licenses. These developments conclusively show that, whatever the (dubious) rationale for the traditional command-and-control administrative regime in the last century, that rationale no longer has any contemporary relevance.

A modern approach to spectrum policy should abrogate the existing framework and replace it with a system that fosters a robust market in which spectrum rights can be freely traded largely independent of any FCC administrative control. To be sure, any new system of spectrum management will require some transitional periods and mechanisms, but the transitional costs are well worth bearing. Under this market-oriented replacement model, the FCC will still retain a role in spectrum management, albeit a much more limited one, primarily as the operator of a rights-clearinghouse akin to a registrar of deeds. The FCC's role would be reduced to performing a modest "zoning" function where necessitated by clear coordination problems and where necessary, in limited special circumstances, to address special needs such as assuring public safety.

There is a widespread, indeed, almost unanimous, consensus among economists and policy experts that a market system that allows flexible use of spectrum is the best way to achieve the most efficient allocation of most resources. This fundamental proposition holds true for the spectrum resource as well – just as Nobel Laureate Ronald Coase contended in 1959 in his famous article, *The Federal Communications Commission*.[136] Coase showed that clearly delineated property rights and market forces, not government control, would lead to the most efficient allocation of the spectrum resource.[137] Again, during the past two decades,

---

136 R. H. Coase, *The Federal Communications Commission*, 2 J. L. & Econ. 1 (1959).

137 As Thomas Hazlett explains, just a year after publication of Coase's *FCC* work, in *The Problem of Social Cost* Coase "demonstrated that resources would be efficiently allocated if rights to property were well-defined and the cost of

Congress and the FCC have taken some steps towards implementing a more market-oriented regime. Now, as Congress considers updating the Communications Act, any new policy framework must clearly support and foster a market in spectrum property rights, which relies on unencumbered auctions to make the initial assignments of spectrum rights, which thereafter should be freely tradable in the secondary market.

II. Spectrum Regulation in the 20th Century

The FCC's role in regulating spectrum in the 20th century was very different from what it should be today. Our understanding of the spectrum throughout much of the last century, when its uses were more rudimentary and based on far simpler, less efficient technologies, was limited compared to today. To be sure, early on some core uses constituted essential services – most notably ship-to-ship and ship-to-shore radio and other radio transmissions needed for emergency services. In this setting, the Federal Radio Commission, beginning in 1927, and then the FCC, which succeeded the FRC in 1934, served two primary functions: to allocate spectrum in specified frequency bands for specified service uses, and within such allocated bands, to assign frequencies to particular users and ensure that such users did not interfere with each other and (especially) with essential services.

It should be noted that the rationale for this form of administrative command-and-control regulation was weak even then. Prior to the establishment of the FRC, the Radio Act of 1912 required the Secretary of Commerce to issue licenses as a matter of right. Under this system, and prior to the creation of the FRC, courts were already developing workable common law

---

trading these rights was modest." Thomas W. Hazlett, *Spectrum Tragedies*, 22 Yale J. on Regulation 242, 244 (2005), commenting on, Ronald H. Coase, The *Problem of Social Cost*, 3 J. L. & Econ. 1(1960).

rules to govern rights in spectrum.[138] The transition away from a system that relied on first possession to establish rights to pervasive regulation of the allocation, transfer, and use of spectrum introduced myriad inefficiencies. As noted above, Ronald Coase's study of these inefficiencies led directly to the development of his Nobel-winning work on externalities and his argument that the government should embrace market-like competitive bidding systems (e.g., auctions) to allocate spectrum.[139] While Coase was often dismissed as a heretic in 1959, no one today seriously argues against the superiority of auctions for allocating spectrum. The other alternatives – comparative administrative hearings and lotteries – consume vast resources to locate spectrum in the wrong hands. Comparative hearings present major public choice concerns because they are an open invitation for never-ending opportunities for lobbying and jockeying in efforts to get the FCC to adopt favorable decisional criteria or other advantages impacting the selection.[140]

Whatever the case that existed for regulation of spectrum in the 20th century, today it is far weaker. Decades of experience with the FCC's various regulatory modalities have demonstrated the weaknesses of command-and-control regulatory methods

---

138   See *Tribune Co. v. Oak Leaves Broadcasting Co.* This 1926 Illinois state court decision is reprinted in Cong. Rec. – Senate 215-219 (Dec. 10, 1926). For a good discussion of the *Oak Leaves* case and related common law developments regarding spectrum property rights before the Federal Radio Act was enacted in 1927, see Thomas W. Hazlett, *The Rationality of U.S. Regulation of the Broadcast Spectrum*, 33 J. L. & Econ. 133, 148-152 (1990).

139   R. H. Coase, *The Federal Communications Commission*, 2 J. L. & Econ. 1 (1959).

140   Lotteries avoid some of the public choice problems plaguing comparative hearings. But they suffer from the serious disadvantage that, unlike auctions, they do not allow the functioning of market-based price signals to determine the highest, best use of the spectrum. To the extent that the FCC is not committed to rules that promote unencumbered "clean" auctions – and it certainly hasn't always been so committed – auctions too are subject to public choice concerns as potential bidders lobby to obtain auction rules that favor themselves relative to their rivals.

relative to market mechanisms that rest on property rights and voluntary contract to allocate spectrum use. Indeed, the correct choices are even more critical today because recent technological advances – particularly the development of spread-spectrum, digital, cellular technology – make it possible to coordinate spectrum use on highly efficient platforms that are relatively resilient to interference. The more intensive use of underutilized spectrum thus increases the overall carrying capacity of the system, which could prove critical for the data-intensive uses of the future.

A consequence of these advances is that spectrum applications have been undergoing convergence. Earlier, broadcast radio and television, satellite communications, point-to-point microwave, and cellular technologies were developed using unique analog encoding and transmission technologies. Today they all increasingly rely on similar technologies to transmit digital data. Just as there is broad consensus that the silos created by Titles II, III, and VI of the Communications Act are obsolete today, disparate treatment of different spectrum applications within the FCC or between the FCC and other agencies are obsolete and should be phased out.

Over roughly the past 30 years, since the advent of the first cellular and digital technologies, the FCC and Congress increasingly have turned to liberalized and flexible-use licenses, and market-like auctions to allocate them.[141] This welcome development should be prologue to a modernized spectrum policy: the Communications Act review and update is an opportunity to complete and normalize the transition away from the 20th century command-and-control spectrum policy to a market-oriented 21st

---

141   It is important to observe, however, that even when the FCC has implemented auctions to award spectrum frequencies, it too often has encumbered the auctions with various conditions designed to tilt the auction results one way or the other or to favor certain bidders, say, by use of bidding credits or requiring certain modes of operation. Encumbering auctions with conditions obviously compromises the market-based price-setting function of the auctions.

century policy premised on flexible use and freely transferable spectrum rights.

Here are some concrete applications of the general theory.

III. Changing Understandings of Spectrum

As in other areas of telecommunications regulation, the convergence of once disparate end uses has largely been driven by digitalization. An ever increasing amount of content, now well over half, is now encoded and transmitted in digital form. As technology continues to advance, the movement toward ubiquitous digitalization – and, hence, toward further convergence – almost certainly will gain momentum.

That technological convergence to digital will make each band of spectrum a closer substitute to all the others for facilitating the transmission of all forms of information. Convergence increasingly allows spectrum initially tasked to one purpose (e.g., distributing television signals) to be converted to other purposes for which it is equally suited (e.g., wireless data transmission). Of course, sometimes the physical attributes relating to specific bands of spectrum (e.g., how it propagates) may render certain bands more or less suited to specific tasks – concerns that are discussed below. But by and large, technological convergence of spectrum use will lead to greater substitutability across the entire spectrum. The new legal regime should be flexible enough to respond to unanticipated technical changes without having to undertake case-by-case reallocation of spectrum across different uses.

In other words, just as "data is data is data," increasingly "spectrum is spectrum is spectrum." While spectrum may never be fully commoditized, across a broad range of present and future applications, including the majority of actual current uses, spectrum is largely fungible. It is hard to justify any substantial regulation of near-commodity goods.

Over time the amount of spectrum needed for any given application has decreased dramatically. In part this is due to digitalization and increasingly sophisticated compression algorithms. The technology that we use to transmit signals is also far more precise, resulting in less signal "bleed" from one band into adjacent bands. For example, spread spectrum technology yields signals that are less subject to interference from other sources and that are less likely to cause interference to other sources.

These technological changes mean that any given application requires substantially less spectrum today than was required just ten or twenty years ago. Of course, this doesn't mean that spectrum is any less "scarce" today than it was then, given that new uses will create heavy demands for spectrum released from traditional uses. Taken as a whole, these changes only mean spectrum use today is much "thicker" and more robust than just a few years ago: at least in principle, many more buyers should be able to participate in the market, each seeking much smaller spectrum allocations than ever before. This is particularly true when we consider "over the top" applications – applications such as Internet-based radio stations, which "transmit" their signals as data over another data network, without requiring a separate spectrum allocation.

The changes over the past 100 years are likely to be exceeded by changes over the next 100 years. Spectrum policy for the 21st century should enable the next century's worth of development on such matters as multicast and broadcast-like content distribution over cellular networks; advances in MIMO (multiple-input and multiple-output) and related technologies, which already are increasing the capacity of existing networks by orders of magnitude; and the latest research in the use of millimeter-wave frequencies for high-speed data transmission.

Similarly, the increasingly powerful "software-defined radio" technologies promise major improvements, especially when combined with wideband radios and variable-frequency oscillators. Historically, each piece of equipment had to be designed to specific applications and frequencies. Newer technologies, however, allow receivers to be manufactured to operate on a very wide range of frequencies – the specific frequencies to be used are selected by software at run-time. For regulatory purposes, this one key change has the potential to sever the historic ties between frequency allocation, licensed uses, and hardware.[142]

IV. A 21st Century FCC Spectrum Policy

Even with these changing circumstances, there remains a role for the FCC to play in a 21st century spectrum policy. Broadly speaking, the FCC should transition from its historical command-and-control administrative role of allocating spectrum for specific uses and assigning frequencies to specific users to a regime that facilitates a free market that lets spectrum users purchase, mortgage, lease, and share spectrum in accordance with their own business plans. This transition is akin to that of the westward expansion of the United States: in the early years, land was given away in large allocations, to create incentives for private parties to develop new uses and unlock natural resources. Once the land had been mapped, land ownership and use thereafter could be coordinated by the market. Similarly, the first century of spectrum policy allocated broad swaths of spectrum to individual uses and users to develop new technologies and resources; in its second century, there is no need to remain fixated on "high-frequency expansion" to map out the basic contours of the usable spectrum. Today's urgent need is to organize an orderly transition so that the FCC's role will be akin to a "recorder of deeds" that facilitates market transactions.

142　The point of discussing some of the current and anticipated technological advances in the last two paragraphs is not to suggest that policymakers should be engineers and understand their intricacies. Rather the point is to show that these advances are consistent with – indeed, strongly bolster the case for – adoption of the flexible use, market-based regime proposed in these comments.

Making this *transition* requires some specific changes to the FCC's current approach to spectrum regulation. Some of these will be of limited duration and will give way as robust markets displace the existing command-and-control regime.

First, as suggested above, the convergence in use through convergent technologies militates in favor of parallel structural convergence. Within the FCC, this means consolidating bureau functions relating to spectrum (e.g., from the Media, Wireless, and International Bureaus). Beyond the FCC, it means rationalizing the spectrum management functions of the Commission and NTIA.

Thus, excepting special circumstances, relating, for example, to national security or public safety, the government should yield voluntarily its control over the spectrum resource, except perhaps for that spectrum that it wishes to retain for its own use. For spectrum that it wishes to retain, the government should have to pay market prices so that, for greater transparency, its own costs are on-budget. But, even here, the government should consider leasing spectrum from the private market to satisfy its service needs. The private sector has greater expertise in developing and deploying wireless networks. Competitive forces – which are increased with federal, state, and local governments as customers – are more likely to ensure those networks are technologically current and economically efficient. Supporting government users' specific needs could lead to development of new consumer-benefitting technologies. They could also exploit important economies of scale that would benefit both consumer and government users.

Second, while there may remain some limited role in rare cases for the Commission to zone spectrum for specific uses which depend on distinctive technical characteristics of transmission, the government otherwise should not hold back spectrum for general uses unless it is prepared to bid in the market against others. And the FCC should move away from regulator-defined protective "guard" bands. Where a given use requires the "extra" protection of a guard band that effectively encumbers adjacent spectrum, the licensee should internalize that cost.

Third, the Commission's approach to unlicensed spectrum needs substantially more theoretical development. While our presumption favors licensed spectrum to promote economic development, an emerging consensus now thinks spectrum policy should embrace both licensed and unlicensed spectrum uses. But the understanding of how that insight is put into operation today, or in the near future, is underdeveloped. So-called unlicensed spectrum today is not actually "unlicensed": its use requires compliance with various requirements imposed by the FCC and industry standards groups. The purpose of these requirements is to mitigate interference and manage scarcity. Users of unlicensed spectrum incur the costs of complying with these requirements and, even when they comply, they still face the risk of congestion.

Importantly, the unlicensed model faces an open question over what happens should its users face scarcity, which could happen if the approach is so widely adopted that its technological approach to scarcity mitigation fails. Should this point of saturation not be reached, licensed and unlicensed spectrum will start to resemble each other: both offer users exclusive use of spectrum; the boundaries of that spectrum are merely being delineated along different dimensions in exchange for either a license fee or the cost of technological compliance. If unlicensed spectrum becomes so popular that its users face scarcity, it is unclear how the Commission would or should respond to this new tragedy of the commons.[143]

---

143   As discussed in the FCC's 2002 Spectrum Policy Task Force Report, there is
an "important caveat" to unlicensed models. In that report, the Commission
observed that: "An important caveat must accompany any recommendation for
a commons model: although there are indications that technology can go a long
way to forestall scarcity concerns, if scarcity eventually does arise in particular
spectrum bands in the future, then the commons model may need to evolve to
address the problem." FCC Spectrum Policy Task Force, *Report of the Spectrum
Rights and Responsibilities Working Group* (November 15, 2002).

Fourth, a proper 21st century spectrum policy will account for the role of receiver design. This discussion results in large part from the social losses that stemmed from the FCC's revocation of LightSquared's permit to deploy a terrestrial data network on spectrum adjacent to that used by GPS systems. The issue also ties into the earlier discussion of software-defined radio and issues about congestion on unlicensed spectrum. In the context of the LightSquared matter it is worth stressing that GPS receivers have long been manufactured on the assumption that spectrum adjacent to that used by the GPS system would only be used, if at all, for low-power satellite applications. By the same token, at no point did the GPS users secure any restrictive covenant or other restriction that would confine LightSquared to limited uses on its own spectrum before LightSquared had acquired the spectrum and spent substantial sums developing it.[144]

The general point is that, in a world of readily-transferred flexible-use licenses, hardware manufacturers' decisions about receiver design should not be allowed to hamper or block the otherwise efficient operation of spectrum markets. Software-defined radio and related technologies increasingly give manufacturers the ability to design their receivers in a versatile way so that they are not tied to specific frequencies, encoding schemes, or protocols. A 21st century free market-oriented spectrum policy such as that proposed here would have encouraged the prompt resolution of the LightSquared – GPS receiver issue in a way that would have mitigated the sizeable social losses that have been incurred as a result of the FCC remaining mired in the throes of the traditional administrative fiat regime. As the LightSquared situation illustrates, the current FCC licensing regime discourages market-based transactions that would allow spectrum to be put to productive use in a way that enhances overall consumer welfare.

---

144   On the LightSquared – GPS interference situation, see Thomas W. Hazlett, *Tragedy of the Regulatory Commons: LightSquared and the Missing Spectrum Rights*, 13 Duke Law and Technology Review 1 (2014).

V. Conclusion

As the Committee moves forward with its review and update process, especially regarding spectrum policy, we urge it to carefully consider and implement the views expressed in this Response.

Chapter 3
Summary of the Third White Paper

"Competition Policy and the Role of the Federal Communications Commission"

"Competition Policy and the Role of the Federal Communications Commission" was the third white paper issued by the House Committee on Energy and Commerce as part of its plan to examine and update the Communications Act. The paper offered brief background on the technological and competitive changes in the market, including those that have given rise to new and competing communications services platforms, with increasing convergence toward IP-based systems. It also recognized the changes in consumer habits in light of those changes. The paper juxtaposed dramatic and rapid market changes with the Communications Act's framework, under which each title to the Act governs a specific sector of the communications economy with disparate regulatory approaches to complementary and potentially substitutable services. And it solicited public comment in response to a series of questions specific to competition policy and the FCC's role in its formulation and application.

Link: http://energycommerce.house.gov/sites/
republicans.energycommerce.house.gov/files/analysis/
CommActUpdate/20140519WhitePaper-Competition.pdf

Response to Questions in the Third White Paper

"Competition Policy and the Role of the Federal Communications Commission"

by

Randolph J. May, President, The Free State Foundation

Seth L. Cooper, Adjunct Senior Fellow, The Free State Foundation

and

Members of the Free State Foundation's Board of Academic Advisors:

Richard A. Epstein, New York University Law School

Justin (Gus) Hurwitz, University of Nebraska College of Law

Daniel Lyons, Boston College Law School

Bruce M. Owen, Stanford University

James B. Speta, Northwestern University School of Law

Christopher S. Yoo, University of Pennsylvania Law School*

before the

Committee on Energy and Commerce

U.S. House of Representatives

June 13, 2014

---

* While the signatories to this Response are in general agreement with the views expressed in these comments, their participation as signatories should not necessarily be taken as agreement on every aspect of the submission. The views expressed should not be attributed to the institutions with which the signatories are identified. This Response is reproduced here in its original form, save for only minor non-substantive edits.

I. Introduction and Summary

Once more we commend the Committee for undertaking this effort to review and update the Communications Act. As we have stressed in our prior Responses to the Committee, this reassessment is necessary because the Communications Act needs updating.

We agree with the Committee's characterization in its Third White Paper that takes proper account of both the technological advances and dramatic marketplace changes. In much the same language used in the Free State Foundation's First Response to the Committee, the Third White Paper explains:

> The evolution of technology from analog to
> digital and narrowband to broadband has
> brought about the integration of voice, video, and
> data services across multiple platforms employing
> various technologies. The ongoing shift away
> from single-purpose technologies toward
> Internet Protocol packet-switching has rapidly
> called into question the adequacy of the current
> Communications Act and the monopolistic
> assumptions on which it is based.[145]

This statement is an accurate characterization of the profound transformation that has occurred in the communications marketplace. As the White Paper states, it is against this backdrop that "an examination of competition policy and the Communications Act is warranted as part of its ongoing update efforts."[146] In order to enhance overall consumer welfare, a new Digital Age Communications Act must be crafted in a way that requires the FCC to

---

145 "Competition Policy and the Role of the Federal Communications Commission" ("Third White Paper"), House Commerce Committee, at 1.

146 Third White Paper, at 2.

take into account the existence of the increasing cross-platform, facilities-based intermodal competition that characterizes the digital environment. The Committee's Third White Paper presents a number of specific and overlapping questions on competition policy. The tenor of the questions makes it clear that the Committee is especially interested, as it should be, in the role that the existence of intermodal competition should play in assessing overall market competitiveness and in formulating regulatory policy.

The generalized framework presented in this response will offer a holistic response to these separate but interrelated questions. This approach fits with our central theme that facilities-based, cross-platform intermodal competition, enabled by the rise of digital and Internet Protocol-based services, has yet to be sufficiently taken into account by the FCC in its decision-making. While new technologies continue to emerge and older technologies evolve in unpredictable ways, at present the communications marketplace is impacted positively by competition among cable firms, telephone companies, satellite operators, fiber providers, and various sorts of wireless companies, each employing their own facilities. In order to encourage the further development of intermodal platform competition on a long-run sustainable basis, the Commission must avoid adopting policies that, in effect, seek to "manage" competition through resale and sharing mandates. What is needed in its place is a consistent, principled competition policy framework premised on facilitating free entry and exit as the basic rule, which should then be qualified by targeted *ex post* remedies rather than by prescriptive *ex ante* regulation.

Stated otherwise, a combination of rapid technological innovation, consumer choice, and disruptive changes in the communications market has altered forever the traditional competitive landscape. These profound structural and technological changes point to the need for a competition policy that leaves free from government regulation those market processes that continue to propel further innovation and competition for new services. Regulatory intervention is only warranted in instances where there is convincing evidence of a market failure that is likely to harm consumers. Absent such evidence of market failure, service and product suppliers should be free to exercise their informed business judgment in an entrepreneurial fashion. Their success will be shaped by how an ever more sophisticated generation of telecommunications consumers respond to their business offers. The interaction of both sides of the market place will outperform any effort by the FCC to chart through government design the direction of future innovations in the ever larger and more complex Internet marketplace.

To this end, under a revised Communications Act, FCC oversight of the modern communications marketplace should be conducted pursuant to a consumer welfare-based standard that relies heavily on antitrust-like microeconomic analysis. That is, the FCC's competition policy should be oriented toward the economically productive and efficient processes by which market participants bring innovative products and services to consumers and respond to changing consumer demands, rather than to any preconceived notions by government officials concerning the shape of the market or the terms and conditions under which services may be offered. From an institutional standpoint, the FCC's competition policy should be geared much more toward *ex post* adjudications than broad *ex ante* prescriptive rulemakings.

II. Competition Policy and Processes for a New Communications Act

While a new Communications Act should not direct the FCC to apply current antitrust precedents in a rigid fashion, it should require that FCC competition policy draw upon the insights of antitrust jurisprudence for purposes of analyzing what kinds of market practices pose competitive issues. As the Free State

Foundation scholars stated in their First Response, adherence to these antitrust-like jurisprudential principles would properly require the FCC to engage in a rigorous economic analysis of market conduct that focuses on actual and potential competitive effects of various firm practices, technologies, and innovations. Such analysis would necessarily take into account the impact of the dynamism – and the "creative destruction"[147] – that characterizes the digital marketplace.

Regulatory prohibitions and sanctions under the new Communications Act should generally be accomplished through focused adjudicatory proceedings. The filing of individual complaints, whether by consumers or market rivals, should contain specific allegations of abuse of market power. The burden should rest on complainants to demonstrate the need for regulatory intervention by clear and convincing evidence of anticompetitive conduct and its likely resulting harm. Any regulatory intervention by the FCC should thus normally be tied to a finding of a threat of market power abuse and a concomitant threat of consumer harm. Furthermore, due to the dynamism that characterizes the modern communications marketplace, these allegations of market failure should show more than some transitory failure that can be met by targeted responses of other market participants. Therefore, any allegations of market failure should be "non-transitory" in order to trigger a Commission response.[148]

Adoption of a competition policy based on a consumer welfare standard grounded in antitrust-like principles necessarily means discarding the indeterminate public interest standard. As already explained further in FSF's Response to Questions in the First White Paper,[149] the current public interest standard confers

---

147   See Joseph A. Schumpeter, *Capitalism, Socialism and Democracy* 87 (3d ed. 1950).

148   See Randolph J. May and James B. Speta, "Digital Age Communications Act," Proposal of the Regulatory framework Working Group, Progress & Freedom Foundation, June 2005.

149   Free State Foundation Response to Questions in the First White Paper,

almost unbridled discretion on the agency without sufficient direction from Congress.[150] The public interest standard is a vestige of monopoly-era assumptions that unwisely assume regulatory intervention as the norm. As we explain below with a few specific examples, this traditional approach places high hurdles to obtaining deregulatory relief even when market conditions have introduced effective competition. Under a revised Communications Act, competition policy should place the burden on the FCC to demonstrate the necessity of regulatory intervention to address market power concerns that threaten harm to consumers.[151]

Application of a marketplace competition standard would make it easier for communications companies to develop ideas and bring new products to market without first having to gain government approval. An *ex ante* regulatory regime that operates mainly through rulemaking inhibits spontaneous innovation and investment by imposing heavy entry barriers on new technologies. Under such a regime, entrepreneurs may feel compelled to submit new services or products to the Commission for review or face the threat of subsequent litigation and sanctions over their lawfulness. An *ex post* process, operating under a proper competition standard, would encourage businesses to bring new services and products to the marketplace without seeking prior regulatory approval.

Establishing a regulatory construct for the FCC favoring *ex post* adjudications necessarily means transforming the FCC into more of an enforcement agency that operates much more like the Federal Trade Commission, at least with regard to competition

---

"Modernizing the Communications Act" (January 31, 2014).

150   Randolph J. May, *The Public Interest Standard: Is It Too Indeterminate to Be Constitutional?* 53 Fed. Comm. L. J. 427 (2001).

151   See Randolph J. May, "A Modest Proposal for FCC Regulatory Reform: Making Forbearance and Regulatory Review Decisions More Deregulatory," *Perspectives from FSF Scholars*, Vol. 6, No. 10 (April 7, 2011); Randolph J. May, "The FCC's Net Neutrality Proposal: The Wrong Way to Use Regulatory Presumptions," Free State Foundation Blog, June 4, 2014.

issues. This transformation does not mean that the FCC necessarily should be precluded from adopting generic rules that define, in advance, certain specific acts or practices that constitute threats of abuse of market power because they cause consumer harm. But such rulemaking authority should be circumscribed by incorporating as a precondition for adoption of a new rule the market failure and consumer harm analysis discussed above.

To be sure, there are some specific but limited areas where the FCC may be granted express rulemaking authority. For example, the FCC should have carefully delineated authority to address interconnection practices that might pose significant consumer harm if the agency finds that marketplace competition is not adequately protecting consumers.[152] This authority is peculiarly appropriate because hold-up problems can easily arise in complex settings that only function well when all carriers, regardless of size or content, have to gain unqualified access to all users of the Internet. Spectrum provides another example where the FCC rulemaking authority may be needed to address interference issues or other technical matters. It should be stressed, however, that the same basic consumer welfare and antitrust-like competition principles should inform the FCC's exercise of its rulemaking authority in these areas. As explained in FSF's Response to the Second White Paper,[153] spectrum policy should transition from a command-and-control model to a property rights-based approach. Consistent with this paradigm shift, FCC spectrum

---

152 See note 149 supra. Parties to an interconnection dispute should be required to engage in some form of dispute resolution process such as mediation prior to seeking FCC decisional intervention. And if it proves necessary for the Commission to intervene to resolve the dispute, the agency should avoid employing traditional administrative public utility-like proceedings in favor of more efficient processes such as baseball-style arbitration. See Randolph J. May, "Testimony of Randolph J. May, President, Free State Foundation," Hearing on "Evolution of Wired Communications Networks," Subcommittee on Communications and Technology (October 23, 2013).

153 "Free State Foundation Response to Questions in the Second White Paper," (April 25, 2014).

policy should emphasize flexibility that allows service providers to respond to marketplace changes without having to endure onerous government processes used to reallocate spectrum across different uses.

Any FCC rules based on competition policy should sunset automatically after an appropriate period of time, say, five years. However, the FCC could be allowed to extend such rules if it affirmatively finds, based on clear and convincing evidence, that there is a market failure that necessitates continuation of the rules to prevent consumer harm.

III. Intermodal Competition: Policy and Process Implications

Under a new Communications Act, FCC competition policy and agency processes should comport with the realities of increasing facilities-based intermodal competition across digital platforms, and they should promote the continued development of facilities-based competition. Too often in the past, for example during the Commission's years-long Unbundled Network Elements ("UNE") proceedings, the agency adopted regulations requiring various forms of network unbundling and facilities sharing. This has been done with the notion that such mandated sharing increases competition, but it generally doesn't accomplish this purpose. Instead, such policies necessitate the existence of an ongoing regulatory program in which the government sets the rates, terms, and conditions under which the unbundled and shared services must be offered. When the required unbundling is excessive, or the regulated sharing price set too low, the new entrant is able to game the system by purchasing elements at bargain rates. Yet if the rates are set too high, the new entrant can resort to market alternatives. FCC policies must guard against the creation of these free options. Yet at the same time, with respect to unique essential facilities, it is critical not to set rates in ways that block new entry.

Many of these issues surfaced in *AT&T v. Iowa Utilities Board*, where the Supreme Court reviewed the FCC's implementation of the network unbundling requirement in the Telecommunications Act of 1996.[154] In invalidating the Commission's UNE rules, the Court concluded that the agency had interpreted the statutory unbundling standard so loosely that it wrongly gave the sharing beneficiaries "blanket access" to the incumbent carriers' networks.[155] Justice Breyer's separate opinion emphasized the ultimate harm to competition caused by the FCC's rules requiring excessive sharing:

> Increased sharing does not by itself automatically mean increased competition. It is in the unshared, not the shared, portions of the enterprise that meaningful competition would likely emerge. Rules that force firms to share every resource or element of a business would create not competition, but pervasive regulation, for the regulators, not the marketplace, would set the relevant terms.[156]

After the Supreme Court's rebuke in *Iowa Utilities Board*, the FCC tinkered with its network unbundling rules before the revised version came back before the D.C. Circuit for review. In *U.S. Telecom Ass'n v. FCC*,[157] the UNE rules were once again invalidated for requiring excessive sharing. As Judge Williams explained, referring to Justice Breyer's *Iowa Utilities Board* opinion, "each unbundling of an element imposes costs of its own, spreading the disincentive to invest in innovation and creating complex issues of managing shared facilities."[158]

---

154   525 U.S. 326 (1999).
155   Id. at 390.
156   Id. at 428 – 429.
157   290 F. 3d 415 (D.C. Cir. 2002).
158   Id. at 427.

All too often, the FCC has failed to grasp this fundamental point. When revising the Communications Act, the goal, as Justice Breyer put it, must be to foster "meaningful competition," not to unwisely maintain "pervasive regulation." And this requires observance of a proper competition standard, such as we have suggested, that favors investment in new facilities over mandated sharing of existing facilities. Under such a proper standard, older technologies can be adapted to new purchases. It was commonly thought as late as 1996 with the passage of the Telecommunications Act that local exchange carriers would be able to maintain a bottleneck position for the foreseeable future. Within a few years, it became clear that cellphone technology, VoIP, and the Internet could provide viable alternatives. Regulatory policy will always go down the wrong path if it ignores the dynamic forces that constantly undercut the creation and maintenance of services monopolies.[159]

Still tied to the silo structure mindset that subjects various services to disparate regulatory requirements, the FCC to date has shown too little interest in evaluating intermodal competition. This lack of interest is perhaps most pronounced when it comes to the substitutability of wireless services for wireline in relation to the overall competitive dynamics of cross-platform rivalry. The FCC has declined to undertake any meaningful analysis of intermodal competition between wireless service and wireline in its *Wireless Competition Reports*.[160] Its *Qwest-Phoenix MSA Order* (2010) and subsequent forbearance orders effectively have rejected cross-platform competition from wireless voice services by imposing a heavy presumption against the substitutability of wireless for wireline.[161] This despite the significant and predictable

---

159   See, for an early statement of this position, Harold Demsetz, *Why Regulate Utilities?*, 11 J.L. Econ. 55 (1968); Richard A. Posner, *Natural Monopoly and Its Regulation*, 21 Stan. L. Rev. 548 (1999).

160   See Seth L. Cooper "Convergent Market Calls for Serious Intermodal Competition Assessments," *Perspectives from FSF Scholars*, Vol. 8, No. 12 (May 2, 2013).

161   See Memorandum Opinion and Order, *In the Matter of Petition of Qwest*

observable losses in wireline market share to wireless. It is striking that during the first half of 2013, 39.4% of households did not have a landline telephone but did have at least one wireless phone.[162] Just 17 years after passage of the 1996 Telecommunications Act, the FCC's *Local Telephone Competition Report* states that, as of December 2013, the number of wireless subscriptions – 305 million – is now *more than three times* the number of wireline access lines – 96 million.[163]

Similarly, in its *Video Competition Report*s the FCC continues to disregard online video as a cross-platform competitive substitute for multi-channel video programming distributor (MVPD) services – even as Netflix has mushroomed into the nation's largest distributor of video programming with over 33 million U.S. subscribers, more subscribers than Comcast and the two satellite TV distributors have.[164] Indeed, almost 50% of U.S. households now subscribe to Netflix or one of the other leading online video distributors, such as Hulu Plus or Amazon Prime.[165] This discounting of the rapidly growing online video distributor market segment in competitive assessments is unwise. It comes on top of the FCC's continued indifference to intermodal competition from direct broadcast satellite (DBS) providers and

Corporation for Forbearance Pursuant to 47 U.S.C. § 160c) in the Phoenix, Arizona Metropolitan Statistical Area, WC Docket No. 09-135 (June 22, 2010). See also note 147 supra.

162   See Stephen J. Blumberg, Ph.D., and Julian V. Luke, "Wireless Substitution: Early Release of Estimates From the National Health Interview Survey, January-June 2013," Division of Health Interview Statistics, National Center for Health Statistics, Centers for Disease Control and Prevention (released December, 2013).

163   See FCC, *Local Telephone Competition Report* (2013).

164   See Seth L. Cooper, "FCC's Video Report Reveals Disconnect Between Market's Effective Competition and Outdated Regulation," *Perspectives from FSF Scholars*, Vol. 7, No. 25 (September 12, 2012).

165   Janko Roettgers, "Close to Half of All U.S. Households Subscriber to Netflix, Amazon Prime or Hulu Plus, GIGAOM, June 6, 2014, at: http://gigaom.com/2014/06/06/close-to-half-of-all-u-s-households-subscribe-to-netflix-amazon-prime-or-hulu-plus/.

telephone company entrants into the MVPD services market. All told, such multiplatform competition has reduced cable providers' share of the multi-channel video market to 55.7% by the end of 2012, down from approximately 60% in 2010.[166] Yet, the video regulations of the early 1990's were all wrongly premised on the faulty assumption that the market power of cable operators could be maintained for the indefinite future. One consequence of this unsound assumption was a raft of must-carry regulations, program carriage regulations, and video device regulations, all of which impose serious threats to operators' First Amendment rights.

Any new Communications Act should place intermodal competition at the center of the FCC's analysis of market competition. The explanatory power of static market indicators such as market concentration or market share is severely limited when dynamic markets characterized by innovation and disruption are under review.[167] Convergence of services and the emergence of new services resulting from the digital transition are testaments to the persistence of market dynamism. Competition between different communications platforms must inform the product and service market definitions to be used by the FCC as part of its analyses of market power and potential consumer harm, including, of course, the Commission's evaluation of the competitive impacts of mergers and other transactions that require agency approval.[168] And these intermodal competition considerations

---

166 See FCC, *Fifteenth Video Competition Report*, (2013). See also Seth L. Cooper, "FCC Report Reconfirms the Reality of the Video Market's Competitiveness," *Free State Foundation Blog* (July 25, 2013).

167 See Dennis L. Weisman, "On Market Power and the Power of Markets: A Schumpeterian View of Dynamic Industries," *Perspectives from FSF Scholars*, Vol. 3, No. 5 (2008).

168 There is extensive literature on the need for reform of the FCC's transaction review process. And many scholars have suggested that, in light of the competition reviews undertaken by the Department of Justice or the Federal Trade Commission, the FCC's role should be limited to ensuring that the proposed transaction complies with all existing agency rules. This would eliminate

should be brought to bear in periodic reports on competition in the communications market – presumably through a reconstituted FCC report that combines its annual wireless, video, and other reports.[169]

Our central point is that the rise of intermodal competition dismantles the analytical underpinnings of the FCC's silo approach to communications services as a whole. As more fully explained in FSF's Response to Questions in the First White Paper, the various silos – whether denominated "telecommunications," "information services," "cable service," "mobile service," or so on – are primarily based on "techno-functional" constructs that do not comport with the realities of digital age technologies and service offerings and the way in which consumers perceive the choices available to them in the marketplace.[170] Technological transitions to all-digital and to all-IP services have furthered the integration and interchangeability of voice, video, and data services regarded as discrete and separate. Consumer expectations for a consistent interface and end-user experience across multiple platforms dictate the end of the prevailing "silo" approach.[171]

Convergence in spectrum applications, described in FSF's Response to Questions in the Second White Paper, offers yet another instance in which the silos created by the Communications Act's Titles II, III, and VI have become increasingly obsolete. Promoting intermodal competition among different spectrum-based

---

the substantial duplication of effort that currently occurs when a proposed transaction is reviewed by both the FCC and the antitrust authorities.

169   See Randolph J. May, "Testimony of Randolph J. May, President, Free State Foundation," Hearing on "Evolution of Wired Communications Networks," Subcommittee on Communications and Technology (October 23, 2013).

170   See note 149 supra.

171   As briefly summarized in the above section and addressed more fully in the Response to the First Paper, because consumer protection issues such as privacy and data security are from part of the FTC's institutional expertise, the FCC should surrender its jurisdiction over such issues to the FTC. See Randolph J. May and Seth L. Cooper, "Any New Privacy Regime Should Mean An End To FCC Privacy Powers," *Perspectives from FSF Scholars*, Vol. 7, No. 9 (April 5, 2012).

applications requires a reoriented analysis that is cut free from the legacy definitional constructs. A revised Communications Act should facilitate a vigorous competition policy that fosters entrepreneurialism by facilitating a flexible use, market-oriented regime. This market-based spectrum regime will allow spectrum resources to move easily to their highest and most valuable use, while simultaneously encouraging the development of new services and products.

IV. Conclusion

As the Committee moves forward with its review and update process, including the evaluation of competition policy, we urge it to carefully consider and implement the views expressed in this Response.

Chapter 4
Summary of the Fourth White Paper

"Network Interconnection"

"Network Interconnection" was the fourth white paper issued by the House Committee on Energy and Commerce as part of its plan to examine and update the Communications Act. The paper began with a brief background of the federal policy regarding interconnection of telecommunications networks as well as its largely non-interventionist approach to Internet Protocol (IP) interconnection. It succinctly identified the rise of Voice of Internet Protocol (VoIP) and Internet data exchange as critical indicators of interconnection evolution. The paper outlined ways in which existing regimes regarding interconnection vary according to the platform or technology subject to regulation. And it solicited public comment in response to a series of questions regarding interconnection, specifically including agreements and the FCC's role in the negotiation and enforcement of such contracts.

Link: http://energycommerce.house.gov/sites/
republicans.energycommerce.house.gov/files/analysis/
CommActUpdate/20140715WhitePaper-Interconnection.pdf

Response to Questions in the Fourth White Paper

"Network Interconnection"

by

Randolph J. May, President, The Free State Foundation

Seth L. Cooper, Senior Fellow, The Free State Foundation

and

Members of the Free State Foundation's Board of Academic Advisors:

Richard A. Epstein, New York University Law School

Justin (Gus) Hurwitz, University of Nebraska College of Law

Daniel Lyons, Boston College Law School

Bruce M. Owen, Stanford University

James B. Speta, Northwestern University School of Law

Christopher S. Yoo, University of Pennsylvania Law School*

before the

Committee on Energy and Commerce, U.S. House of Representatives

August 8, 2014

---

* While the signatories to this Response are in general agreement with the views expressed in these comments, their participation as signatories should not necessarily be taken as agreement on every aspect of the submission. The views expressed should not be attributed to the institutions with which the signatories are identified. This Response is reproduced here in its original form, save for only minor non-substantive edits.

I. Introduction and Summary

Once again, we commend the Committee for undertaking its sustained effort to review and update the increasingly anachronistic Communications Act. This update is not only timely but necessary, given the rapid rate of technological change, and the concomitant change in communications and information services markets, since Congress last updated the law with the Telecommunications Act of 1996.

We also commend the Committee for using the Fourth White Paper to focus specifically on interconnection. We agree with the Committee's recognition that the interconnection of communications networks "has been at the heart of communications policy" for a century, and further, that it should be an integral component of any Communications Act update.[172] As twentieth-century communications networks give way to the all-IP-based networks of the future, there is still a useful role for a government regulator to play in overseeing the interconnection of the various privately-operated networks that comprise the nation's communications infrastructure.[173]

But going forward, this role should be noticeably different – presumptively less interventionist – in scope than it is under the current Act. That conclusion is consistent with the transition to more competitive communications and information services markets. Rather than overseeing enforcement of a general duty to interconnect, as the current Act requires, the law should presume that interconnection agreements between IP-based networks will be negotiated on a voluntary basis, as they have been throughout

---

172 "Network Interconnection" ("Fourth White Paper"), House Commerce Committee, at 1.

173 See Free State Foundation Response to Questions in the First White Paper, "Modernizing the Communications Act" at 13 (January 31, 2014); Free State Foundation Response to Questions in the Third White Paper, "Competition Policy and the Role of the Federal Communications Commission" at 7-8 (June 13, 2014).

the Internet's history with minimal disruption. The Commission should intervene only upon a finding that denial of interconnection poses a substantial, non-transitory risk to consumer welfare, and that marketplace competition is inadequate to correct the problem. And in those rare instances when intervention is necessary, the Commission should solve the impasse by using some form of dispute resolution mechanism, such as mediation or some form of arbitration, rather than by resorting to current rate case-like adjudicatory procedures. This revised interconnection mandate is consistent with our view of the FCC's future role, not as regulator of monopolistic common carriers subject to public utility obligations, but rather as a sector-specific competition authority protecting consumer welfare in a competitive and dynamic marketplace.

II. Modern Interconnection Markets Are Competitive and Dynamic

As the Fourth White Paper notes, the Telecommunications Act of 1996 sought to bring a "pro-competitive and deregulatory framework" to local telephone markets, in part by loosening the reins on existing interconnection obligations.[174] Section 251 placed a general duty on all telecommunications carriers to interconnect with one another,[175] and it imposed additional duties on legacy incumbent local exchange carriers (ILECs) to negotiate in good faith, to connect at any technically feasible point in its network, to provide a level of service equivalent to what it delivers to itself, and on reasonable, nondiscriminatory terms.[176] The law allowed networks to freely negotiate the terms of these interconnection agreements, but it subjected ILEC agreements to review by state regulators, and mandated compulsory arbitration by state

---

174   Fourth White Paper at 1.

175   47 U.S.C. § 251(a).

176   *Id.* § 251(c).

regulators in the event of an impasse.[177] This interconnection duty and the concomitant arbitration and review procedures were part of the Act's broader movement from a heavily-regulated local telephone monopoly to a regime of "managed competition" within largely intrastate local telephone markets.

While the 1996 Act represented progress toward a deregu-lated communications marketplace, since that time market de-velopments have obliterated the line between local and long-dis-tance service that in 1996 justified state-level involvement in interconnection negotiations. Indeed, contrary to the dominant expectations of the day, the wireline telephone sector as a whole has receded dramatically, replaced by wireless networks (which are subject only to Section 251(a)'s general duty to interconnect) and by IP-based networks. Because the Commission has classified IP networks as "information services" rather than "telecommuni-cations services," they are not subject to the Act's interconnection duties.

Interconnection has flourished despite—or perhaps more accurately, because of—this unregulated environment. Com-mentators often describe the Internet, accurately, as a "network of networks." Interconnection agreements stitch this network to-gether. The IP interconnection market is a "collection of 35 thou-sand autonomous systems bargaining with one another through arms-length transactions" to shuttle traffic among the Internet's end-points.[178] As one might expect, these agreements inevitably contain wide variations in the terms under which parties inter-connect and exchange traffic with one another. Interconnection agreements can run hundreds of pages, governing a wide range of conditions, and they are typically covered by non-disclosure agreements that reflect the competitively sensitive nature of those terms. All are freely negotiated on a voluntary basis, without a

---

177   *Id.* § 252.

178   Christopher S. Yoo, The Dynamic Internet: How Technology, Users, and Businesses Are Transforming the Network at 55 (2012).

regulator-enforced duty to interconnect or government review of an agreement's terms.

The interconnection market is diverse, in part because of the flexibility of voluntary negotiations to respond to changing market conditions. The two dominant forms of interconnection are peering (in which two networks agree to provide reciprocal access to each other's end-user consumers) and transit service (in which one network agrees to provide access to all Internet destinations). Many peering agreements between networks of comparable size are on a settlement-free basis, though the market has seen a rise in "paid peering" when traffic flows disproportionately in one direction. Transit is usually sold on a volume basis, with the sending network paying the receiving network to deliver its traffic to its destination. To avoid being dependent on one interconnection agreement or network, many content providers and transit networks sign interconnection agreements with multiple networks to route traffic to end-users, a practice known as "multi-homing."[179] The ability to forge these multiple pathways blunts the holdout power that any one carrier might have over the system, so that it is no longer appropriate to presume, as was done in 1996, that multiple parties had no choice in the pathway over which they transmitted their traffic. Competitive conditions, in a word, have been improved markedly by these advances in technology.

The interconnection market is also dynamic and evolving in response to changing patterns of Internet-based consumption. For example, the rise of Internet-based video services such as Netflix and Hulu has increased the volume and the vector of traffic flowing over IP networks, thus spawning alternatives to traditional peering and transit models. Content Delivery Networks such as Akamai maintain a distributed network of servers around the country, which store local copies of their clients' content for delivery to consumers. Because the content traverses fewer interconnections, CDNs can be a high-quality, low-cost

179 *Id.* at 62-64.

alternative to traditional transit for streaming video and other similarly-situated content providers. Some high-volume content providers have also begun to engage in self-provision, building their own server farms to store their content. This allows them to interconnect directly with end-user broadband networks rather than relying on transit providers for delivery. These innovations help provide lower-cost, higher-quality service for applications that need more than the "best efforts" delivery that marks the traditional public Internet. And as the "Internet of Things" of all manner of connected computing devices grows, one can imagine an entire class of network-based services that could survive on less-than-best-efforts service.[180] As traffic flows change, interconnection agreements negotiated on a voluntary basis have the flexibility to adapt to suit.

The underlying technology of IP interconnection is also more complicated than in prior communications network architectures. The transition from circuit-switched TDM networks to packet-switched statistically-multiplexed networks – the transition that has enabled much of what is described above – greatly increases the number of factors that interconnection must address. Previously, interconnection meant physically connecting two networks and providing a relatively simple method for allocating a circuit between endpoints for the duration of a phone call.

Today, interconnection still involves a physical connection between networks. But the algorithmic logic governing how that connection is used must make real-time routing decisions on a per-packet level, sometimes factoring in information about current network conditions; it must respond in real-time to configuration changes across the network; it must incorporate real-time

---

180   The "Internet of Things" generally refers to a wide variety of connected devices with embedded computing capabilities, such as smart thermostats, heart monitoring implants, biochip transponders on farm animals, automobiles with built-in sensors, and field operation devices that assist fire-fighters in search and rescue.

resource allocation logic; it must respond to congestion events; it must have logic for buffering and sorting packets as they arrive at switches and routers. Importantly, many of these factors conflict: Making a network more resilient to congestion, for instance, can increase latency and jitter when congestion does occur. And there is no "one-size-fits-all" configuration that works well for all uses or users. In other words, interconnection among IP networks is precisely the sort of relationship that is best negotiated on a case-by-case basis by parties that understand their particular needs and how those needs best map onto the underlying technology.

Finally, IP interconnection markets are competitive. Content providers have multiple options to deliver their content to consumers—from transit providers with nearly global footprints to regional providers that rely on interconnection agreements to route traffic onward to consumers.[181] Some provide transit service only, while others provide complementary services as well. Although pricing schedules are often protected by nondisclosure agreements, there is a general consensus that competition has driven down Internet transit prices continuously and precipitously each year since the modern Internet's inception in the 1990s. Interconnection consultant William Norton calculates, based on informal surveys, that the average per-Mbps price for generic non-commit transit service has fallen from roughly $1200 in 1998 to $12 in 2008 and $0.94 in 2014—an average rate of decline of over 30 percent each year.[182] TeleGeography similarly estimates that transit prices have fallen 26% annually from 2007 to 2012,[183]

---

181   See Dan Rayburn, *How Transit Works, What it Costs & Why It's So Important*, Feb. 24, 2014, available at http://blog.streamingmedia.com/2014/02/transit-works-costs-important.html.

182   William B. Norton, The Internet Peering Playbook: Connecting to the Core of the Internet at 34 (2013).

183   See TeleGeography Press Release, *IP Transit Prices Steepen*, Aug. 2, 2012, available at http://www.telegeography.com/products/commsupdate/articles/2012/08/02/ip-transit-price-declines-steepen/.

and Streaming Media Analyst Dan Rayburn has noticed similar trends in CDN prices.[184]

III. The FCC's Limited, but Important, Role in Interconnection

The key to the tremendous growth and complexity of IP-based networks has been the supplanting of a public utility regime by a free-market oriented regulatory model. Given the importance of interconnection to the healthy functioning of any communications system, we recognize that the Commission should continue to play a role in overseeing interconnection between network providers. But rather than the heavy-handed regulator of the 1934 Act or the competition "manager" of the 1996 Act, the future Commission's interconnection authority should be circumscribed, and instances of actual intervention should be rare.

In our prior responses, we have noted that a future Commission should play a role with regard to interconnection. In light of a residual holdout problem, perhaps its role should be above and beyond the general role we envision for the Commission as enforcer of a sector-specific competition standard grounded in antitrust principles.[185] Our position is informed by many of the concerns that animated the Digital Age Communications Act Working Group's proposal in 2005.[186] First, the Supreme Court's *Trinko* decision leaves some uncertainty regarding whether one can order interconnection under an antitrust-based unfair

---

184   See Dan Rayburn, *The State of the CDN Market*, May 2014, available at http://www.streamingmedia.com/dansblog/2014CDNSummit-Rayburn.pdf.

185   See, e.g., Free State Foundation Response to Questions in the First White Paper, "Modernizing the Communications Act" at 13 (January 31, 2014).

186   See Randolph J. May and James B. Speta, "Digital Age Communications Act," Proposal of the Regulatory Framework Working Group, Progress & Freedom Foundation, June 2005 ("DACA Working Group"). As before, we acknowledge the debt owed to the DACA Working Group, within which many of these proposals originated.

competition standard.[187] Second, denial of interconnection can sometimes be a rational economic strategy whereby a single network can attempt to dominate a market in a way that harms consumers.[188] Finally, there are many non-economic social benefits to a unified communications network as an inclusive forum for news, education, free expression, access to emergency services, and facilitation of democratic self-government. The Commission has a role to play in safeguarding these positive externalities of increased interconnection.

But while interconnection mandates can sometimes correct market failures in ways that enhance social welfare, this gain must be balanced against regulation's own potential inefficiencies. As Michael Katz and Carl Shapiro have noted, private institutions such as new players or standards-setting bodies may arise to achieve coordination and internalize the externality without government intervention.[189] Second, government intervention may unreasonably favor the status-quo, ossifying current practices at the cost of blocking or raising costs on innovative emerging technologies.[190] Third, regulators often lack the information needed to determine which course maximizes total surplus.[191]

Given these potential risks, we recommend that the Committee reject a general duty to interconnect with other IP-based networks. As an empirical matter, there appears no need to impose such a duty: content and application providers have a plethora of options available to deliver their content to consumers, and

187   See Verizon Communications, Inc. v. Law Offices of Curtis V. Trinko, LLP, 540 U.S. 398, 415 (2004); Philip J. Weiser, *Goldwasser, The Telecom Act, and Reflections on Antitrust Remedies*, 55 Admin. L.J. 1(2003); James B. Speta, *Antitrust and Local Competition under the Telecommunications Act*, 71 Antitrust L.J. 99 (2003). But see United States v. Terminal R.R. Ass'n, 224 U.S. 383 (1912); Otter Tail Power Co. v. United States, 410 U.S. 366 (1972).

188   See DACA Working Group at 26.

189   *Id.* at 26; see Michael L. Katz & Carl Shapiro, *System Competition and Network Effects*, 8 J. Econ. Persp. 93, 112-13 (1994).

190   Katz & Shapiro at 112-13.

191   *Id.*

through multi-homing, they often leverage multiple options simultaneously to reduce the risk that any one network can exercise market power against them. There have been very few instances in which interconnection disputes have been brought to the Commission's attention, and those have largely been solved through private negotiations.[192] Moreover, given the wide range of potential interconnection options, from peering (free or paid) or transit to CDNs, self-provision, and more, the contours of such a duty would be difficult to define and enforce. Attempts to do so could inadvertently lock-in existing practices and reduce the market's ability to respond dynamically to changing trends in content and application markets.

Accordingly, we recommend that the Commission be given the authority to intervene to address only those specific interconnection practices that pose a substantial and non-transitory risk to consumer welfare.[193] The consumer welfare focus assures that the Commission's decision whether to intervene is based upon an explicit finding that the practice poses harm to competition generally, rather than to one specific competitor. The Commission's authority should further be premised upon an explicit finding that marketplace competition is insufficient to protect consumer welfare.[194]

In those (hopefully rare) instances warranting intervention, the Commission should mimic private dispute resolution mechanisms rather than take on, in a new context, its traditional role as a public utility regulator. Typically, it should condition its intervention on a requirement that the parties first submit their dispute to mediation. If mediation is unsuccessful, the Commission should devise some form of arbitration process, perhaps, for example,

---

192 See Randolph J. May, "Testimony of Randolph J. May, President, Free State Foundation," Hearing on "Evolution of Wired Communications Networks," Subcommittee on Communications and Technology (October 23, 2013).

193 DACA Working Group at 24.

194 *Id.* at 24-25.

some form of "baseball-style arbitration," in which each side submits a proposed "last and best offer" and the arbitrator then chooses one.[195] This approach limits the Commission's discretion to interfere in ongoing interconnection disputes and places the burden primarily on the parties themselves to find a solution to the impasse. There may be other dispute resolution models that ought to be considered as well. And, whatever mechanism is chosen, the process must be such that there will not be undue delay in delaying resolution of the dispute.

We also strongly recommend the Committee reject calls by some to require public disclosure of the terms of interconnection agreements. Mandating the detailed disclosure of specific, confidential business-to-business agreements negotiated between sophisticated parties in a highly competitive market is likely to do more harm than good to competition. One reason is that any distinctive structure of these agreements could well contain important trade secrets whose value is lost if made public. A second reason is that, as the industrial organization literature emphasizes, the sharing of competitively sensitive information among rivals can facilitate tacit collusion on price. For this reason, the Supreme Court, antitrust authorities, and the Commission itself have long stressed that disclosure of pricing and cost information can be harmful to competition, especially in markets like telecommunications that involve significant barriers to entry.[196]

Finally, state authorities should play a much more limited role with regard to interconnection than they do under the now obsolete 1996 Act regime. This is consistent with the evolution of telecommunications markets and the now almost-complete elimination of intrastate "local" markets as a competitively important

195　See Randolph J. May, "Testimony of Randolph J. May, President, Free State Foundation," Hearing on "Evolution of Wired Communications Networks," Subcommittee on Communications and Technology (October 23, 2013).

196　See Daniel A. Lyons, *Compelled Disclosure of Internet Interconnection Agreements Creates Anticompetitive Risks*, Free State Foundation *Perspectives,* Vol. 9, No. 22 (2014).

classification. Today's information networks are largely national in scope, and neither providers nor customers easily distinguish between interstate and intrastate communications. Attempts by state regulators to review or interfere with national interconnection agreements can have the unintended consequence of balkanizing the nation's information infrastructure and of compromising the economies of scale generated by interstate operations. As noted in our response to the First White Paper, state regulators should retain a prominent voice with regard to consumer protection issues.[197] And any state should have standing to bring an interconnection dispute to the Commission's attention if the dispute adversely harms that state's constituents. But the Commission should have final decisionmaking authority regarding whether to intervene, given that it normally is in a better position to assess the costs and benefits of intervention from a national scope.

IV. Conclusion

As the Committee moves forward with its review and update process, we urge it to carefully consider and implement the views expressed in this Response, as well as the previous Free State Foundation Responses. We look forward to continuing to play a constructive role in this process leading to a much-needed update of the Communications Act.

---

197   See Free State Foundation Response to Questions in the First White Paper, "Modernizing the Communications Act" at 18 (January 31, 2014).

Chapter 5
Summary of the Fifth White Paper

"Universal Service Policy and the Role of the Federal Communications Commission"

"Universal Service Policy and the Role of the Federal Communications Commission" was the fifth white paper issued by the House Committee on Energy and Commerce as part of its plan to examine and update the Communications Act. The paper outlined the statutory and non-statutory principles upon which federal universal service policy is premised. It then described the four major areas of universal service program funding. And it solicited public comment in response to a series of questions specific to universal service policy and the FCC's role in its formulation and application.

Link: http://energycommerce.house.gov/sites/
republicans.energycommerce.house.gov/files/analysis/
CommActUpdate/20140822White%20Paper-USF.pdf

Response to Questions in the Fifth White Paper

"Universal Service Policy and the Role of the Federal Communications Commission"

by

Randolph J. May, President, The Free State Foundation

Seth L. Cooper, Senior Fellow, The Free State Foundation

and

Members of the Free State Foundation's Board of Academic Advisors:

Richard A. Epstein, New York University Law School

Justin (Gus) Hurwitz, University of Nebraska College of Law

Daniel Lyons, Boston College Law School

Glen O. Robinson, University of Virginia Law School

James B. Speta, Northwestern University School of Law

Christopher S. Yoo, University of Pennsylvania Law School*

before the

Committee on Energy and Commerce, U.S. House of Representatives

September 19, 2014

---

* While the signatories to this Response are in general agreement with the views expressed in these comments, their participation as signatories should not necessarily be taken as agreement on every aspect of the submission. The views expressed should not be attributed to the institutions with which the signatories are identified. This Response is reproduced here in its original form, save for only minor non-substantive edits.

I. Introduction and Summary

Once again, we commend the Committee for undertaking this effort to review and update the Communications Act. And we also commend the Committee for using the Fifth White Paper to focus on universal service reform. As the Committee correctly noted, "[t]he principle of universal service has been at the heart of federal and state telephone policy" nearly from its inception.[198] And rightfully so.

The universal service principle, which supports access to basic communications service for all Americans, is an important component of the social safety net. Keeping all members of society connected, regardless of income or social status, redounds to the benefit of those who can afford to pay as well as those who cannot afford to pay for access to the network. And, in so doing, this helps reduce America's digital divide, which, parenthetically, we observe is also narrowed by adoption of efficiency measures that reduce the costs of access for all.

But while universal service is a laudatory goal, its execution has often left much to be desired. In the nearly two decades since the system was overhauled in the Telecommunications Act of 1996, Universal Service Fund expenditures have doubled in size to $8.3 billion annually.[199] Yet much of that new funding has been spent on projects that have little to do with the traditional goals of universal service. The Federal Communications Commission has admitted that the current system distorts investment and competition in myriad ways, while the Government Accountability Office pointedly has criticized the program's lack of oversight and accountability.

At the same time, advances in technology that have eroded

---

198   "Universal Service Policy and the Role of the Federal Communications Commission" ("Fifth White Paper"), House Commerce Committee, at 1.

199   USAC 2013 Annual Report at 35, available at http://usac.org/_res/documents/about/pdf/annual-reports/usac-annual-report-Interactive-Layout-2013.pdf.

legacy jurisdictional boundaries and service distinctions mean that the fund is drawing from a shrinking base of interstate and international telecommunications revenue. The increased USF payouts, coupled with the shrinking revenue base, has caused the USF "surcharge" paid by consumers to skyrocket from 3% in 1998 to 16.1% in the fourth quarter of 2014.[200] This 16.1% surcharge necessarily operates as a tax on all interstate and international calls. It is counterproductive to fund universal service by taxing the very same services the Commission seeks to promote.

To the Commission's credit, it has recently begun to reform the most troubling portions of the current Universal Service program, and it has begun to shift the program's focus from supporting ordinary telephone to supporting broadband service, recognizing the importance of high-speed Internet access to modern society. But these reforms, while not insignificant, represent only the first steps in what should be a complete reform-minded overhaul of the program. As part of its Communications Act Update process, the Committee should refocus the Universal Service program on what ought to be its core mission: (1) assisting consumers who cannot afford broadband access, through market-based, consumer-empowering initiatives that befit an increasingly competitive broadband marketplace; and (2) supporting build-outs only in unserved areas and then only through economically efficient mechanisms such as reverse auctions (whereby firms bid the lowest rate that they will accept to supply narrowly targeted subsidy support.) In addition, any Lifeline program that is maintained for those who cannot otherwise afford service must also be operated in a way that incorporates means to prevent fraud and waste.

Moreover, the Commission should abandon the antiquated contribution mechanism currently in place, with the present USF surcharge set at a 16.1% tax on all interstate and international

---

200   See Michael O'Rielly, *Disturbing Trend in USF Spending*, Official FCC Blog, July 7, 2014, available at http://usac.org/_res/documents/about/pdf/annual-reports/usac-annual-report-Interactive-Layout-2013.pdf.

calls. Instead, it should avoid within-sector distortions by funding subsidies through appropriations from the general treasury where they are subject to more intensive scrutiny. This approach would allow Congress to set a hard budgetary cap on expenditures for a definite period, which, in turn, will encourage the Commission to wring inefficiencies out of the system in order to live within the congressional funding restraints. It would also make today's hidden tax more transparent in a way that provides for greater accountability to Congress and taxpayers.

II. The Troubled Universal Service Program at a Crossroads

The basic tenet of universal service – that the government should assist those who cannot afford basic access to the telecommunications network or those who live in unserved areas – has been a cornerstone of telecommunications policy for nearly a century. In economic terms, this assistance is justified by network effects: the larger the number of people a network reaches, the more valuable that network is to each user. A public policy to maximize telecommunications subscribership levels benefits not only those who cannot otherwise afford access, but all other subscribers as well. Universal service also helps maximize the utility of the network for society as a whole, by improving civic participation levels, economic opportunities, and public safety.

Unfortunately, however, less than a quarter of the Universal Service Fund's $8.3 billion annually goes toward Lifeline and Link Up, the two programs that directly serve this laudable goal.[201] Over $2.2 billion in 2013 went to E-Rate, a program that provides computers and broadband access to schools and libraries, which should be funded through ordinary education budgets, usually at the state and local level.[202] Congress, the GAO, and the

---

201 See 2013 Annual Report, supra note 199, at 35.
202 *Id.*

Commission's own Office of the Inspector General have criticized E-Rate for lax oversight and the potential for abuse, including purchasing wasted resources or resources far exceeding the beneficiary's reasonable needs.

Even this $2.2 billion is dwarfed by the $4.2 billion – half of all USF contributions – spent on the Universal Service program's High-Cost Fund, which directly subsidizes the operations of telephone companies in rural areas and other so-called "high-cost" markets.[203] Ostensibly, the High-Cost Fund is designed to defray the costs of wiring and serving customers in unconventional, expensive service areas. But in practice, the fund is a wasteful form of corporate welfare that distorts competition and inhibits cost reduction in those areas it claims to serve.

The High-Cost Fund routinely subsidizes service in many areas where subsidies are unnecessary or harmful. Worse, it often has perverse distributional effects, by subsidizing wealthy rural consumers, who are never subject to means-testing, but who can easily afford to pay the full cost of access. For example, economist Thomas Hazlett notes that in 2005, residents in the Jackson Hole, Wyoming, service area received $282 per subscriber in annual subsidies, even though their incomes and net worth were well above the national average.[204]

The High-Cost Fund also encourages inefficient investment, a fact the Commission has readily acknowledged. For rate-of-return carriers, which consume over $2 billion annually in subsidies despite serving less than five percent of all telephone users, subsidies are calculated based upon the carrier's "embedded costs." Perversely, the higher the firm's costs, the more it will receive in subsidies. There are few mechanisms in place to prevent "gold plating," that is, the installation of equipment far more expensive than necessary to meet a community's needs, solely to increase

---

203   *Id.*
204   Thomas Hazlett, *"Universal Service" Telephone Subsidies: What Does $7 Billion Buy?* at 3 (2006).

the company's rate base. Until recently, there was also no cap on per-line support, which allowed Hawaii's Sandwich Isles Communications Company to receive over $10,000 per line per year in subsidies between 2005 and 2010.[205] Similarly, Beaver Creek Telephone Company received over $454,000 in 2008 to serve just 27 lines, which amounted to a cost of almost $17,000 per line.[206] Other companies receive thousands per line in annual subsidies, often in areas that wireless providers can serve at a fraction of the cost.[207]

Finally, the High-Cost Fund lacks sufficient oversight and accountability. And the government is the first to admit this shortcoming. The GAO released a scathing report in June 2008 entitled "FCC Needs to Improve Performance Management and Strengthen Oversight of the High-Cost Program."[208] Indeed, the Commission's own audits concluded that in 2006-07, the fund made "erroneous" payments totaling $970 million, or almost 25 percent of all high-cost spending.[209] The assessment concluded that the fund was "at risk" under the Improper Payments Information Act of 2002, which labels any federal program risky if its error rate exceeds $10 million or 2.5 percent of annual disbursements.

To its credit, the Commission has sought in recent years to reform various Universal Service programs while maintaining fidelity to its view of its statutory mandate. For example, its 2011 order placed a "firm budget" on the High-Cost Fund, holding

---

205   See http://transition.fcc.gov/wcb/iatd/monitor.html; Federal Communications Commission Response to United States House of Representatives Committee on Energy and Commerce Universal Service Fund Data Request of April 1, 2009.

206   Id.

207   See Hazlett, supra note 7, at 20 (Noting that Nextel offered wireless service throughout the Sandwich Isles Communications Company service area, and that even satellite phone service can be less expensive than the per-line amounts paid to some carriers.).

208   GAO 08-633, 2008 WLNR 13168534 (July 14, 2008).

209   Office of Inspector General, Federal Communications Commission, The High Cost Program Initial Statistical Analysis of Data from the 2007/2008 Compliance Attestation Examinations, November 2008.

HCF spending to the 2011 level of $4.5 billion – though it falls short of the "hard cap" that many commenters recommended, as the Commission, unfortunately, has committed only to "an automatic review trigger" if annual claims against the fund threaten to exceed that amount.[210] The Commission also has capped the amount of annual assistance that a carrier may receive to $3000 per line. But the handful of carriers that currently receive assistance above the capped amount may petition the Commission to waive the cap under standards that seem unclear and indefinite. These mild limitations are welcome first steps toward arresting the growth of a fund that has been spiraling out of control for nearly a decade. But they are half measures at best.

More monumentally, the Commission began reorienting the fund's focus from telephone service to broadband access. Price-cap carriers receiving High-Cost Fund assistance for voice telephony must also offer their customers fixed broadband service at specific speeds, while rate-of-return carriers must make such service available upon a customer's reasonable request. The Commission has also established a new Connect America Fund to bring broadband service to underserved areas. In 2012, it followed this initiative with a pilot program to test whether it is feasible to extend Lifeline support to include broadband access as well as traditional telephone service.

Importantly, the Commission's order also signaled a sea change in the agency's philosophy for administering the Universal Service program. The Commission has explicitly and repeatedly endorsed a "more incentive-based, market-driven approach" to distributing subsidies. For example, the 2011 order eliminates High-Cost subsidies to any carrier whose service area is already

---

210   See Connect America Fund, Report and Order and Further Notice of Proposed Rulemaking, FCC 11-161 (Nov. 18, 2011), at ¶ 563 (explaining that in the event that HCF demand exceeds $4.5 billion in a year, the Wireline Bureau will "provide to the Commissioners a recommendation and specific action plan to immediately bring expenditures back to no more than $4.5 billion.").

served by an unsubsidized provider. Any such aid distorts competition without furthering the goal of providing telecommunications services to those who do not have reasonable access. In areas where two or more subsidized providers are competing, assistance will be based on each carrier's costs rather than the arcane and wasteful "identical support rule" that paid new providers on the basis of the incumbent's legacy cost structure. And most significantly, the Commission has endorsed reverse auctions as the ultimate mechanism to administer CAF subsidies so that providers will compete for subsidies by bidding to provide predetermined service throughout a defined service area at the lowest cost.

III. Pursuing More Fundamental Universal Service Reforms

The Commission's recent reforms respond to two significant changes in the telecommunications marketplace. The first is the shift from monopoly to competition. The industry is no longer dominated by monolithic legacy telephone companies that agreed to serve everyone in a service area in exchange for state protection from competition. Today, over 99 percent of U.S. census tracts are served by two or more fixed broadband providers at the benchmark speed of 3 Mbps downstream and 768 kbps upstream, and that number remains at 82 percent even at the higher 10 Mbps/1.5 Mbps threshold. These numbers increase to 99 percent and 98 percent respectively if wireless broadband is included.[211] The Commission's endorsement of incentive-based, market-driven policies recognizes that the fund should capitalize on this shift by embracing competition to aid the unserved.

The second is the shift from circuit-switched voice telephony to IP-based broadband networks. The Commission is absolutely

---

211 See Federal Communications Commission, Internet Access Services: Status as of June 30, 2013, at 9 fig.5(a), 10 fig.5(b) (2014), *available at* https://apps.fcc.gov/edocs_public/attachmatch/DOC-327829A1.pdf.

right that high-speed Internet access will be the primary telecommunications network of the future, with voice service representing one of many applications available to consumers. The Communications Act recognizes that universal service is an evolving level of telecommunications service, and high-speed Internet access is increasingly required to participate in modern economic and civic life.

Yet the Commission's efforts do not reflect the magnitude of reforms demanded by these fundamental industry shifts. The Commission seems content to focus on individual components of the universal service program by modifying existing programs used to support the old telephone industry to fit today's very different broadband market. This piecemeal approach squanders major opportunities presented by the migration to broadband by importing defective programs in the current system into a dynamic new marketplace. The correct mission is to design new and better subsidy programs that are tailored to meet the unique challenges that broadband presents.

The Commission's reforms are also hobbled by statutory language that has allowed the USF program to stray far from its core mandate of assisting those who genuinely cannot afford access to the network. The 1996 Act's extension of subsidies to high-cost carriers, libraries, schools, and rural health care facilities drove both the monumental growth in fund expenditures and the well-deserved complaints about waste and corruption in the fund's administration. In the process, it created a host of new recipients whose voices are shaping the reform debate. It is no surprise that the 2011 reforms adopted many elements of the so-called "ABC Plan," an earlier comprehensive proposal backed by a consortium of subsidy-receiving telephone companies. While it's rational that these companies would seek a voice in reform negotiations, their concern with preserving existing subsidies is ultimately in tension with efforts to achieve more fundamental change.

America's migration to broadband networks presents a

once-in-a-generation opportunity to transform an outdated, mismanaged subsidy program. Congress should capitalize on this window of opportunity by adopting market-based initiatives with the objective of assisting low-income persons who cannot afford basic broadband Internet access and ensuring that any subsidies directed to service providers are narrowly targeted only to unserved geographic areas and then are provided only in the most economically efficient manner. And support for low income persons should be provided directly to them, rather than to supporting carriers as their proxies in ways that distort broadband competition. Finally, these efforts should be undertaken in a financially responsible manner, minimizing costs and funding those costs in a way that does not overly burden broadband consumers and providers.

A. Reinventing Lifeline

On the subsidy side, we recommend that Congress maintain a reformed Lifeline subsidy program to assist eligible low-income persons. The Lifeline program might consist of a voucher program similar to a telecommunications version of the food stamp program, or a fund-provided broadband phone card. Eligibility should be determined by the means-testing currently undertaken for Lifeline eligibility: proof of income below an appropriate income level, or participation in one of many other means-testing programs such as Medicaid, food stamps, SSI, or the Section 8 Housing Program. States could adopt alternative eligibility criteria with Commission approval, which would help ensure the program is flexible enough to meet each state's unique needs.

The voucher would be designed to provide "basic broadband service" to eligible recipients. Numerous advocates have pushed the Commission to define broadband service in terms of minimum speeds—often defined as speeds sufficient to stream high-quality video. But the current universal service program does not fund video service, which we consider outside the core of basic social services to which all should have access. To avoid this ambiguous mission creep, Congress should define "basic broadband service" by listing those essential services that the program should facilitate online, such as including emergency assistance,[212] voice service, access to government services and information, and perhaps basic e-commerce. The Commission should then define the minimum speed necessary for a broadband connection that provides reliable access to those services that Congress deems essential.

Like Lifeline today, the voucher would allow eligible consumers to purchase basic broadband service for a Commission-defined subsidized rate. Eligible consumers would receive a nontransferable voucher equal to the average market rate for basic broadband service in the service area, minus the subsidized rate. In service areas where one broadband provider has market power, the Commission could prevent price-gouging by limiting the voucher to an amount sufficient to assure a reasonable rate of return to a reasonably efficient provider in that service area.

But the voucher itself would be set at a fixed, portable amount that the consumer could take to any participating telecommunications provider. Any provider that accepts a voucher must agree to provide basic broadband service to voucher holders throughout the service area at no more than the subsidized rate plus the voucher amount. The consumer need not use the voucher to purchase basic broadband service; he or she could instead choose to apply the voucher to receive (presumably less expensive) voice-only service, or as a credit toward a suite of more advanced telecommunications services, if the provider makes

---

212   Many states currently impose high E911 surcharge fees on providers of Lifeline services. This seems counterproductive because one of the reasons for creating a Lifeline service is to help ensure that low-income persons have access to these services. And, in any event, it has been shown that these fees often bear little or no relationship to the cost of providing E911 or other forms of emergency assistance, but rather are imposed as general revenue-raising measures.

these alternatives available to the public generally. This flexibility extends the promise of at least voice access to those eligible households that cannot afford broadband even at the subsidized rate, without locking in voucher recipients to basic broadband if they are willing to pay for additional services.

Perhaps there are other ways to operate an efficient and effective Lifeline program. But a portable voucher structure gives purchasing power directly to low-income individuals, allowing them to participate in the telecommunications marketplace like any other consumer. And it allows the fund to benefit directly from competition among broadband providers. To attract recipients and avoid customer defection, providers must compete on price and service as they do in the marketplace generally. Moreover, because the voucher amount depends upon the average market price for broadband service, less efficient providers have economic incentives to improve their operations while hyper-efficient competitors are rewarded accordingly. Furthermore, the vouchers are technologically neutral: any provider willing to offer basic broadband service would be eligible to participate, regardless of the platform through which the customer is served.

B. Funding Buildouts

Of course, vouchers do not help those areas of the country that are not yet wired for broadband service because extending the network would be cost-prohibitive without some subsidy. For these areas, Congress may wish to maintain the buildout assistance contemplated in the Commission's 2011 order for a limited transition period. Funding, however, should flow only to those areas not currently served by an unsubsidized broadband provider, and recipients should meet stringent buildout requirements and agree to provide service throughout the service area. This aid should be distributed through reverse auctions, and the total amount should be subject to an annual cap that is reduced over time as the market fills in the existing gaps in our broadband network.

C. State Block Grants as an Alternative

Obstacles to universal broadband service vary considerably by geography and are often informed by local knowledge. Given the variation in conditions, it may be wise for state regulators to take the lead going forward. Though universal service is a national goal with positive spillover effects that cross jurisdictions, Congress may consider using block grants to harness this state-level knowledge in pursuit of broader federal goals. Under this proposal, a state could choose to opt out of the federal program, and instead receive a block grant from the Commission equal to the estimated total payout of vouchers to residents of that state. The state would then be free to administer a state-level universal service fund with this money, funding end-users, high-cost carriers, equipment purchases, or whatever else it determines is necessary to overcome key stumbling blocks to furthering universal service in that state. The state's continued funding would be contingent upon its compliance with Commission-determined performance metrics (such as the state's broadband adoption rate) to assure that the state is spending its dollars in ways that help achieve the overall federal objective.

D. Eliminating High-Cost Fund and E-Rate

The primary obstacle to these reforms, as with any broadband universal service proposal, is cost. Broadband access is more expensive than the voice service that the fund currently provides, so any attempt to subsidize broadband service necessarily would increase the size of the fund. But substantial savings may be obtained by phasing out those programs that do not directly serve the goal of bringing access to those who cannot afford it.

Thomas Hazlett notes that much of E-Rate's library and school funding probably replaces other sources of funding that would have provided the same services through other means. This is particularly true in wealthier E-Rate communities such as Beverly Hills, California, and Fairfax County, Virginia, where income is well above the national average.[213] But even setting aside this objection, broadband access is no more integral to a library or school's mission than textbooks or laboratories. The costs of this access should be incorporated into the institution's own budget to be paid the same way as its other expenses, so the local community can make an accurate assessment of whether it needs such service, and, if so, at what level or cost.

Similarly, Congress should commit the Commission to slowly but firmly phasing out the High-Cost Fund assistance given directly to carriers, by steadily decreasing the amount of annual subsidies available until the program is eliminated. This means that in areas with high annual operating costs, the cost of broadband access may rise, perhaps substantially, to reflect the true costs of service. But this is not necessarily a significant problem. Like E-Rate, High-Cost Fund support is not means-tested, so much of it flows to communities where few would drop off the network even if forced to pay full market rates. And the voucher program is targeted to local area rates, which will help mitigate the effect of the price increase on those who are genuinely at risk of cancelling service at an unsubsidized rate.

More fundamentally, the simple fact that a community's telecommunications costs are high is not, alone, a reason to subsidize the community as a whole. There are a wide range of goods integral to modern life whose costs vary dramatically by location—for example, housing, food, or gasoline. Yet few suggest that the higher cost of a two-bedroom apartment in Manhattan compared to Houston means all Big Apple residents should receive a hefty federal housing subsidy. Similarly, the Universal

213   See Hazlett, supra note 204, at 51-52.

Service program should not subsidize what is effectively a lifestyle choice by those who could afford broadband access in most areas but choose instead to live in a region with a high cost of service.

IV. Implementing a New Contribution Regime

Finally, Congress should abandon the fund's increasingly arcane contribution methodology. The existing contribution system relies on anachronistic distinctions between interstate and intrastate service and between telecommunications and information services. Yet both of these are largely irrelevant in the broadband age.[214] There is no reason to preserve a regime that induces consumers and carriers to adopt technology based in part on its regulatory classification rather than its intrinsic value. Moreover, it seems counterproductive to fund universal service by taxing the very services the Commission seeks to promote. As the Federal-State Joint Board noted, "larger USF contributions increase the risk that telecommunications services will become unaffordable for some, or even a substantial number, of consumers."[215] At 16 percent of every bill, the surcharge may already be retarding telecommunications use.

The simplest and most elegant solution to the contribution problem is simply to fund universal service through a line item in the federal budget like most other entitlement programs. Other proposed solutions, such as a tax on telephone numbers or IP addresses, while likely an improvement over the existing revenue-based regime, suffer from similar problems as the current methodology: they are underinclusive and encourage strategic behavior by consumers, while discouraging provision of the very

---

214   High-Cost Universal Service Support, Order on Remand and Report and Order and Further Notice of Proposed Rulemaking, 24 FCC Rcd 6475, 6656 (Nov. 5, 2008).

215   High-Cost Universal Service Support, Federal-State Joint Board on Universal Service Recommended Decision, 22 FCC Rcd 20477, 20483 (Nov. 19, 2007).

services the program seeks to subsidize. Paying for universal service from the general treasury would improve the transparency of the program by vesting oversight in Congress or the Commission rather than the murky, semi-private Universal Service Administrative Company. It would also apply a hard budgetary cap to expenditures established for a defined period of time, requiring the Commission to wring inefficiencies out of the system in order to serve the public within congressional funding restraints. A primary objection to such a shift might be the public's distaste for new entitlement programs. But in reality this approach would not create a new entitlement; rather, it would simply make a hidden tax more transparent, in a way that provides for greater accountability to Congress and taxpayers.

Chapter 6
Summary of the Sixth White Paper

"Regulation of the Market for Video Content and Distribution"

"Regulation of the Market for Video Content and Distribution" was the sixth and final white paper in the series that was issued by the House Committee on Energy and Commerce as part of its plan to examine and update the Communications Act. The paper provided brief background on the history of FCC oversight of broadcast TV, cable video, as well as direct broadcast satellite (DBS) services, and it acknowledged the rise of video services delivered via broadband. It then described various barriers to entry for providing video services, including FCC spectrum licensing as well as state and local government video franchising. Next, the paper included an overview of existing legislative regimes and regulations regarding access by video service providers to video programming, and added a few observations about developments that have rendered the video market substantially different from the last time Congress reviewed the Communications Act. Lastly, the paper solicited public comment in response to a series of questions specific to regulation of the market for video content and distribution.

Link: http://energycommerce.house.gov/sites/
republicans.energycommerce.house.gov/files/analysis/
CommActUpdate/20141210WhitePaper-Video.pdf

Response to Questions in the Sixth White Paper

"Regulation of the Market for Video Content and Distribution"

by

Randolph J. May, President, The Free State Foundation

Seth L. Cooper, Senior Fellow, The Free State Foundation

and

Members of the Free State Foundation's Board of Academic Advisors:

Michelle Connolly, Duke University

Richard A. Epstein, New York University Law School

Justin (Gus) Hurwitz, University of Nebraska College of Law

Daniel Lyons, Boston College Law School

Bruce M. Owen, Stanford University

Glen O. Robinson, University of Virginia Law School

James B. Speta, Northwestern University School of Law

Christopher S. Yoo, University of Pennsylvania Law School*

before the

Committee on Energy and Commerce

U.S. House of Representatives

January 23, 2015

---

* While the signatories to this Response are in general agreement with the views expressed in these comments, their participation as signatories should not necessarily be taken as agreement on every aspect of the submission. The views expressed should not be attributed to the institutions with which the signatories are identified. This Response is reproduced here in its original form, save for only minor non-substantive edits.

Randolph J. May and Seth L. Cooper

I. Introduction and Summary

Once more, we commend the Committee for undertaking its sustained review and update of the increasingly anachronistic Communications Act. As is the case with the other areas that Free State Foundation scholars have addressed in prior Responses, updating the Communications Act on video policy is especially timely. Since Congress passed the Cable Acts of 1984 and 1992 and the Telecommunications Act of 1996, the video service market has been marked by rapid technological changes and increased competitive activity, which undermines the case for direct regulation.

As the next section of this Response amply demonstrates, breakthrough innovation backed by heavy entrepreneurial investment has for the last twenty-five years drastically transformed the video market. These same forces will continue to reshape its landscape going forward. Convergence toward digital and Internet Protocol-based services and cross-platform competition from competing delivery technologies now offer consumers video capabilities and content choices hardly imaginable when Congress last spoke on video policy in the 1990s. The legacy video regulatory apparatus rested on an excessive concern with perceived scarcity, monopolistic power, lack of distribution outlets, and minimal consumer choice. Legacy video regulations based on those outdated concerns now impose disparate regulatory treatments on competing services, which threaten to stifle future innovation. In short, video policy must be reformed to reflect the demonstrably changed market conditions detailed in the body of this Response.

Historically, constitutional permissibility of several aspects of legacy video regulations rested on upon claims of scarcity and monopoly-like conditions in the market. Given the magnitude of the continuous ongoing technological and competitive changes, the case for exempting video services, with minor exceptions, from the standard First Amendment protections given to other forms of speech has evaporated. Standard rule of law norms now impose on Congress an obligation to adopt a new approach to video services that satisfies these First Amendment constraints.

Replacement of the legacy video services regime should be part and parcel of a new Digital Age Communications Act. A new policy framework for video services should rely on the same fundamental principles applicable to other digital services. Bringing video into a single, unified framework for digital services furthers the goals of policy simplicity and harmony.

The Communications Act's ubiquitous public interest standard, under which so much of video regulation takes place, is unfit for a digital age market characterized by cross-platform video competition. That standard is under-protective of free speech involving video content and other editorial decisions by video service providers. The FCC's open-ended authority to regulate "in the public interest" should be largely curtailed.

Transition of video to digital and IP-based platforms has rendered the current silo statutory structure obsolete and inequitable. And most of the distinct prescriptive rules for various types of video services are unsuited to competitive conditions in today's converged digital market for video services. Thus, for example, basic tier cable channel requirements, basic tier cable rate regulations, program carriage mandates, and must-carry and retransmission consent rules should be eliminated, albeit with reasonable transition periods to allow time for adjusting existing arrangements and protecting reliance interests. Regulation of video services would occur through case-by-case adjudications under the same competition-based standard that we articulated in our First White Paper Response. Thus, except in limited circumstances involving public safety, emergency notifications, or the like, regulation of video services, like other services, would be dependent on findings of consumer harm and market power.

Further, the FCC's existing authority over cable subscriber privacy and over DBS subscriber privacy should be transferred to the Federal Trade Commission. Consolidating consumer privacy for video services and other digital services within the FTC's jurisdiction would establish a consistent set of rules for data privacy policy, enforced by a single agency.

This clean-slate approach to video policy is consonant with the dynamic and competitive video services market that now exists and it is consistent as well with First Amendment objectives. To advance both consumer welfare and rule of law norms, consistent with our recommendations regarding other communications services, Congress should integrate video services into a generally applicable framework that presumes regulation is unnecessary absent evidence of consumer harm and market failure.

II. Today's Video Market Is Dynamic and Competitive

Clearly, today's video market is far different from that of the early 1990s, when video programming subscribers enjoyed only a limited number of analog, standard definition channels, little or no interactive capabilities, and fewer pricing or premium content choices. At that time, the cable television industry was deemed "highly concentrated."[216] Most Americans had access to only one multichannel video programming distributor (MVPD), and cable operators controlled approximately 95% of the national market for video programming subscribership.[217] Those conditions prompted Congress to impose regulations intended to facilitate the entry of new programmers and media voices.

Much of the Cable Acts of 1984 and 1992 as well as the FCC's video regulations are based on these analog-era monopolistic assumptions about the video market. Recognizing the competitive, innovative nature of the video market is a necessary first

---

216 1992 Cable Act, § 2(a)(4).

217 *See In re* Annual Assessment of the Status of Competition in the Market for the Delivery of Video Programming, *Further Notice of Inquiry*, 26 FCC Rcd 14091, 14092-3, ¶ 2 (2011).

step in bringing about legislative and regulatory reforms to reflect marketplace realities.

Dramatic advances in technology and market entry by cross-platform competitors have now led to proliferation of video capabilities and content choices for consumers. New technologies, new pricing options, and new sources of distribution now supply evolving consumer demands. In short, rapid innovation backed by heavy entrepreneurial investment has drastically transformed the video market of twenty-five years ago and continues to re-shape its landscape.

Snapshots of market share data often give a poor measure of market competitiveness, when market shares can rapidly change as consumer behavior responds to technologies and services, many of which are introduced by new market entrants. But even in static terms, readily available evidence reveals vibrant compet-itive market conditions with multiple choices for consumers. In the Commission's *Fifteenth Video Competition Report*, the agency found that by the end of 2013, cable providers represented only 55% of the more than 100 million households that subscribe to all multichannel video program distributors (MVPDs).[218] Mean-while, direct broadcast satellite (DBS) providers and "telephone" providers gained market share, claiming about 33.6% and 8.4% of all MVPD subscribers respectively.[219] At the end of 2011, 98.6% of subscribers or 130.7 million households had access to at least three MVPDs, 35.3% or 46.8 million households had access to at least four, and some areas had access to as many as five MVPDs.[220] The Commission's latest report is already a year-and-a-half old. Not surprisingly, trend lines suggest the market

---

218   Annual Assessment of the Status of Competition in the Market for the Delivery of Video Programming *Fifteenth Report*, MB Docket No. 12-203 at ¶ 3 (released July 22, 2013), available at: http://transition.fcc.gov/Daily_Releases/Daily_Busi-ness/2013/db0801/FCC-13-99A1.pdf.

219   *Id.* at ¶ 27.

220   *Id.* at ¶ 36.

share of cable operators has slipped even further relative to DBS providers.[221]

The emergence of telephone MVPDs like AT&T and Verizon marks one of the most significant changes in market competition for MVPD services.[222] U-verse and FiOS services were available to one-third of U.S. homes by the end of 2010, accounting for approximately 6.5 million customers.[223] By the end of June 2012, AT&T's U-verse and Verizon's FiOS services grew to 10.7 million subscribers combined.[224] CenturyLink's Prism TV also continues to expand, boasting approximately 175,000 subscribers at the end of 2013.[225] Within recent weeks, CenturyLink announced first that it reached a franchise agreement with Portland, OR, and second that it is pursuing an agreement in Minneapolis, MN.[226] By year's end 2013, Consolidated Communications Holding's had 110,000 subscribers to its IPTV service.[227] Moreover, entrants such as Google Fiber, certainly not lacking in financial resources, are also supplying consumers with new options,

---

221 *See* SNL Kagan, Press Release: "Multichannel Video Subscription Count Drops by a Quarter Million in 2013 According to Research by SNL Kagan" (March 19, 2014), at http://www.snl.com/InTheMedia.aspx. See also Roger Yu, Pay-TV providers see first yearly customer loss, USA Today (March 19, 2014), at http://www.usatoday.com/story/money/business/2014/03/19/pay-tv-2013-customer-loss/6628769/.

222 Annual Assessment of the Status of Competition in the Market for the Delivery of Video Programming, *Fourteenth Report*, MB Docket No. 07-269, at ¶ 80 (released July 20, 2012) (citing SNL Kagan, *Cable TV Investor: Deals & Finance*, Oct. 31, 2011, at 7-8).

223 See SNL Kagan.

224 *Fifteenth Report*, at ¶ 3.

225 *See* SNL Kagan.

226 *See* Mike Rogoway, "CenturyLink strikes cable TV deal with Portland; service could begin in 2015," *The Oregonian* (Dec. 20, 2014), at http://www.oregonlive.com/silicon-forest/index.ssf/2014/12/centurylink_strikes_cable_tv_d.html; Erin Golden, "CenturyLink looks to enter Minneapolis cable market," *Star Tribune* (Dec. 23, 2014), at http://www.startribune.com/local/minneapolis/286634151.html.

227 *See* SNL Kagan.

employing advanced infrastructure technologies and high-speed functionality.

If anything, it seems likely that the competitive capabilities and potential of former telephone companies in providing MVPD services have been underappreciated. Fiber-to-the-home services such as Verizon FiOS have been lauded for providing reliable video because of the enhanced speed, capacity, and other performance capabilities of their infrastructure. At the same time, moreover, "telephone" company MVPDs continue to upgrade their systems through a network fiber build out, that reaches the neighborhood or network node. Expansion of fiber is being combined with innovative techniques for upgrading the capabilities of VDSL (very-high-bit-rate digital subscriber lines) lines. Such solutions parallel regular cable network systems upgrades, enabling "telephone" MVPDs to achieve speeds that are far more comparable to cable ISPs than to older DSL networks.[228]

Increasingly, consumers have gained access to high-capacity wireless broadband services capable of streaming HD and other video content. According to estimates of wireless mobile broadband network coverage cited in the FCC's *Seventeenth Wireless Competition Report* (2014), as of January 2014, AT&T's coverage extended to 98.7% of the population, Verizon Wireless' to 96.9%, T-Mobile's to 92.1% and Sprint's to 89.8%.[229] Those numbers do not include regional or local wireless broadband providers, such as U.S. Cellular, C-Spire, and nTelos.

Next-generation wireless network upgrades continue to

---

228 *See, e.g.*, Reply Comments of ADTRAN, Inc., Applications of Comcast Corporation, Time Warner, Inc., Charter Communications, Inc., and Spinco To Assign and Transfer Control of FCC Licenses and Other Authorizations, MB Docket No. 14-57 (Nov. 7, 2014), at http://apps.fcc.gov/ecfs/document/view?id=60000979824.

229 *In re* Annual Report and Analysis of Competitive Market Conditions With Respect to Mobile Wireless, Including Commercial Mobile Services, *Seventeenth Report*, WT Docket No. 13-135 (released Dec. 18, 2014)(Chart VI.B.1 – Estimated Mobile Wireless Coverage in the U.S. by Provider, Jan. 2014).

exhibit increased speeds and capacity, making wireless an increasingly viable competitive alternative – indeed a potential substitute for – wireline broadband. For most major wireless broadband providers, average LTE speeds range between 30 and 40 Mbps,[230] enabling a wide range of video viewing functionalities. With regard to video, about half of all broadband consumers access mobile TV apps each month; 17% use mobile TV apps weekly; and 16% use mobile TV apps daily.[231] These numbers are certain to increase, as wireless broadband providers are unveiling new technologies that will enhance wireless video viewing capabilities. For instance, Verizon and AT&T plan to launch their LTE multicast video services in the near future. "LTE Multicast" promises to deliver live TV signals wirelessly to mobile devices more efficiently than unicast delivery, because multiple users can watch the same multicast stream as it is delivered from a single cell site.[232] Sprint is deploying its enhanced LTE service, with peak download speeds of 60 Mbps.[233]

Over-the-air broadcast TV continues to offer consumers a further option, albeit now in HD and with multi-casting

---

230 *See, e.g.*, Lynn La, "4G LTE Showdown: How Fast is Your Carrier?" *CNet* (Aug. 5, 2014), available at: http://www.cnet.com/news/4g-lte-showdown-how-fast-is-your-carrier/.

231 TGD, Press Release: "Half of Adult Broadband Users Now Engage Mobile Video Apps at Least Once a Month" (June23, 2014), available at: http://tdgresearch.com/tdg-half-of-adult-broadband-users-now-engage-mobile-video-apps-at-least-once-a-month/; *Id.* ("39% of adult broadband users engage mobile video apps on a portable computer, compared to 30% who do so using a tablet, and 22% that do so using a smart phone").

232 Jeff Baumgartner, "Verizon CFO: LTE Multicast 'Pivotal' to Mobility," *CNet* (Aug. 12, 2014), available at: http://www.multichannel.com/news/technology/verizon-eyes-2015-lte-multicast-video/382678; http://www.lightreading.com/video/mobile-video/verizons-multicast-lte-video-to-arrive-in-2015/d/d-id/710057; Jeff Baumgartner, "AT&T Plays Ball With LTE Broadcast," *CNet* (Jan. 8, 2015), at http://www.multichannel.com/news/technology/att-plays-ball-lte-broadcast/386750.

233 Sprint, Press Release: "Sprint Spark" (June 24, 2014), available at: http://newsroom.sprint.com/presskits/sprint-spark.htm.

capabilities. While recent estimates of audience size suggest that the number of people who rely exclusively on broadcast TV is down substantially from the early 1990s, today's consumers readily combine broadcast TV viewing with low-cost alternatives such as the broadband-enabled online video distributor ("OVD") services like Netflix.[234] Broadcast TV mobility options are also available.[235]

Indeed, the OVD market segment has also emerged as a disruptive force in the video market. By the end of 2013, Netflix had 44 million customers,[236] compared to 21.7 million Comcast customers and 11.4 million Time Warner Cable customers.[237] OVDs have made their programming available on smartphones, tablets, and laptops as well as new connected devices like Roku, Apple TV, and game consoles like Xbox 360, PS3, and Wii.

A recent survey indicates that more than a third of TV households subscribing to MVPD services also subscribe to Netflix.[238] Those numbers suggest that video consumers value OVDs as a complement or add-on to MVPD services. Individual cable and broadcast networks have also embraced streaming services. CNN, ESPN, MTV, and ABC are among the networks that offer their content by live stream or on-demand.

---

234 *See* Christopher S. Stewart, "Over-the-Air TV Catches Second Wind, Aided by Web," *Wall Street Journal* (Feb. 21, 2012), at http://www.wsj.com/articles/SB1000 1424052970204059804577229451364593094.

235 *See, e.g.*, George Winslow, "CES: RCA to Launch Mobile DTV-Capable Tablet," *Broadcasting&Cable* (Jan. 4, 2013), at http://www.broadcastingcable.com/news/ technology/ces-rca-launch-mobile-dtv-capable-tablet/49701?nopaging=1.

236 Victor Luckerson, "Netflix Mulls Price Tiering as Subscriber Numbers Soar," *Time* (Jan. 22, 2014): http://business.time.com/2014/01/22/ netflix-number-of-subscribers-grows-greatly-in-q4/.

237 William Alden, "The Comcast-Time Warner Deal, by the Numbers," *New York Times* (Feb. 13, 2014): http://dealbook.nytimes.com/2014/02/13/the-comcast-time-warner-deal-by-the-numbers/?_php=true&_type=blogs&_r=0.

238 Wayne Friedman, "DVR Service Soars In U.S.," MediaDailyNews (Jan. 2, 2015), at http://www.mediapost.com/publications/article/241028/dvr-service-soars-in-us.html?edition=79069.

On the other hand, OVDs also fill the role of disruptive competitor to MVPD services. According to one report, during the third quarter of 2014, "homes receiving programming solely through broadband service reached 2.57 million, or 3 percent of total TV households."[239] Also, nearly half of households that do not subscribe to pay-TV services have Netflix. Additional OVD or over-the-top (OTT) services, relying exclusively on broadband connections, are also being launched. These include Dish's Sling TV – limited number of cable channels at discount price – and HBO's direct-to-consumer service.[240] Of course, consumers can also make stand-alone purchases of movies, TV episodes, or TV series, or certain video channels through streaming apps and mobile platforms like Apple's iOS, Google's Android, and Amazon Prime.

Today's video market is also characterized by the ongoing replacement of analog systems with digital; the rapid expansion of high-definition broadcasting and TV ownership; the rise of cloud-based user interfaces, digital video recorder (DVR) options, video-on-demand functions, as well as TV-Everywhere and other mobility capabilities.[241] The Commission's *Fifteenth Video Competition Report* notes that, as of 2012, more than 74% of households have sets capable of receiving digital signals, including

---

239   USTelecom, "Broadband Spurs Increased Digital Video Viewing," *USTelecom Blog* (Dec. 11, 2014), at http://www.ustelecom.org/blog/broadband-spurs-increased-digital-video-viewing.

240   *See, e.g.*, David Carnoy, "Dish's new Sling TV Internet TV service starts at $20, features ESPN, Disney Channel, CNN, TNT, and other channels," *CNET* (Jan. 5, 2015), at http://www.cnet.com/news/dish-launches-20-sling-tv-streaming-video-service-with-channel-lineup-that-includes-espn-disney/; Mike Farrell, "OTT's Real Impact Eludes Nets – and Ops," *Multichannel* (Dec. 15, 2014) at http://www.multichannel.com/news/ott/ott-s-real-impact-eludes-nets-and-ops/386284.

241   George Winslow, "TV Everywhere Video Consumption Doubles," *Broadcasting & Cable* (Dec. 16, 2014), at http://www.broadcastingcable.com/news/technology/tv-everywhere-video-consumption-doubles/136435.

HD signals.[242] 4K ultra-HD TV sets and content are now being marketed to consumers.[243]

Streaming apps and mobile platforms like Apple's iOS and Google's Android provide consumers access to video programming offered by cable operators Comcast, Cox, Charter, Cablevision, and Bright House. Comcast and Time Warner Cable programming can be accessed by Microsoft's Xbox 360. Time Warner Cable provides its subscribers access to video-on-demand services through devices like Roku and Samsung Smart TVs.

Non-cable operators like DirecTV, DISH, AT&T, and Verizon have also made their content available through iOS and Android mobile platforms and devices. DirecTV offers its programming through DIRECT Ready TVs and Samsung TVs. Additionally, DISH and DirecTV offer their programming through smart TVs that do not require a set-top box or other receiver.

This survey is necessarily far from comprehensive. Additional options become available almost weekly, if not daily. But the dramatic technological and competitive advances in video services witnessed in the last quarter-century are abundantly and indisputably documented. Video policy must be reformed to reflect these emphatically changed market conditions.

III. First Amendment Constraints Should Prompt Video Policy Reform and Guide Policy Implementation

As indicated above, a significant number of statutory restrictions on video services and the Commission's regulations of such services were premised on a now primitive 1990s snapshot of the video market. The constitutional permissibility of several

---

242   *Fifteenth Report*, at 5, ¶ 7.
243   Mark Hachman, "We saw lots of 4K TVs at CES. 4K content? Not so much." *TechHive* (Jan. 14, 2015), at http://www.techhive.com/article/2867546/we-saw-lots-of-4k-tvs-at-ces-4k-content-not-so-much.html.

aspects of those laws and regulations similarly rested upon claims of monopoly-like conditions in the market.

Given the magnitude of technological and competitive changes of the last quarter-century, the case for exempting video services from First Amendment protections accorded other forms of speech has evaporated. Rule of law norms impose an obligation on Congress to establish a new approach to video services that satisfies First Amendment standards. In other words, in light of today's video market conditions, conformity to First Amendment requirements should dictate a light-touch approach that covers only real, not imagined, market failures.

In the words of the Supreme Court, "leading First Amendment precedents have established the principle that freedom of speech prohibits the government from telling people what they must say."[244] Courts have recognized First Amendment protections against compelled speech in the context of modern media communications.[245] Yet a variety of existing regulations, in fact, do tell video service providers what they must say, overriding editorial decisions with government proscriptions. For instance, decades-old must-carry regulations require MVPDs to carry broadcast TV content not of their own choosing, curtailing MVPDs' discretion, in their role as publishers, to determine channel lineups and arrange channel tiers.

Program carriage regulations designed to protect video programmers unaffiliated with MVPDs from "discrimination" amount to forced-speech mandates by substituting the government's judgment concerning program channel selection and line-up placement for that of an MVPD. Program access regulations effectively require vertically integrated MVPDs, operating in their role as publishers, to act as speakers in settings not of their own

---

244   *Rumsfeld v. CAIR*, 547 U.S. 47, at 61 (2006).

245   *See, e.g., Miami Herald Publishing Company v. Tornillo*, 418 U.S. 241 (1974) and *Pacific Gas & Electric Company v. Public Utility Commission*, 475 U.S. 1, 9 (1975).

choosing by making their programming available to competitors on terms and conditions that are subject to second-guessing by the FCC. Agency-defined "must-have" categories of programming, such as sports networks, designed for purposes of enforcing program access requirements verge on content-based speech controls. And "leased access" regulations, which require MVPDs, again operating in the role of publishers, to make available certain amounts of their channel capacity to third parties subject to government-set rate controls, deprive MVPDs of editorial control over any video programming on the leased channels.

To preserve its regulatory power, the FCC has for many years relied upon rationales that have always been analytically suspect and are today increasingly factually unsupportable. Most notably, in *Red Lion Broadcasting Co. v. FCC* (1969), and *Turner Broadcasting System, Inc. v. FCC* (1994), the existence of so-called spectrum "scarcity" and cable monopoly "bottlenecks" were asserted to excuse selectively applied regulations restricting the free speech of TV broadcasters and cable providers.[246] Many of those restrictions imposed on cable video services have subsequently been extended to other MVPD services, including satellite video providers. In light of these recent developments, it is not surprising that federal courts have challenged both the scarcity and bottleneck rationales for pervasive regulation of speech in video services media.[247] Congress should take

---

246 *Red Lion*, 395 U.S. 367; *Turner Broadcasting*, 512 U.S. 622.

247 *See Comcast v. FCC*, 579 F.3d 1, 8 (D.C. Cir. 2009) ("the record is replete with evidence of ever increasing competition among video providers…Cable operators, therefore, no longer have the bottleneck power over programming that concerned the Congress in 1992"); *Comcast v. FCC*, 717 F.3d 982, 994 (D.C. Cir. 2013) (Kavanaugh, J., concurring). ("In today's highly competitive market, neither Comcast nor any other video programming distributor possesses market power in the national video programming distribution market"); Fox v. FCC, 613 F.3d 317, 326-327 (2d Cir. 2010) (recognizing dramatic changes in technology and competition but concluding it is bound by U.S. Supreme Court precedent regarding spectrum scarcity); *Time Warner Cable v. Hudson*, 667 F.3d

this disconnect into account when it revises its rules for video services.

Courts typically defer to both Congress and federal agencies whenever public policy depends upon judgments involving marketplace economics. While piecemeal judicial dismantling of the legacy video regulatory apparatus might vindicate free speech rights in today's new video market conditions, it could also introduce unintended policy glitches and uncertainties that a uniform approach to all video market segments and providers could avoid. Congress should be the first mover in establishing a new framework for video services. Moreover, respect for First Amendment rights should lead Congress to avoid regulation that abridges free speech. It should also spur Congress to remove outdated regulatory burdens on speech in light of changed marketplace conditions.

Taking the First Amendment seriously as a policymaking guide means authorizing government intervention in the video market only where such intervention demonstrably serves a compelling government interest while burdening protected speech as little as possible. It also means carefully tailoring such intervention to employ the least restrictive means possible for this limited class of objectives. And it means regularly revisiting such regulatory interventions to ensure that they still serve a compelling government interest, with repeal as the default option. The new policy framework for today's competitive, convergent, digitally-driven media marketplace should provide equal speech protections to all video services, regardless of the media or technology used. Regulation is warranted only after a showing of a demonstrable and

---

630 (5ᵗʰ Cir. 2012) (concluding that the First Amendment prohibits modern speech media from being subject to selective, discriminatory regulations and striking down certain state video franchise requirements). On other occasions, federal circuit courts have concluded that the MVPD market characterized by bottlenecks when the Cable Act of 1992 was passed is now "mixed," with competition varying according to geographic region. *See Cablevision v. FCC*, 597 F.3d 1306, 1314 (D.C. Cir. 2010); *Cablevision v. FCC* 649 F.3d 695, 712 (D.C. Cir. 2011); *Time Warner Cable v. FCC*, 729 F.3d 137 (2d. Cir. 2013).

compelling market failure based on specific technical and market characteristics.[248]

IV. A Market-Power Framework for Video Services in the Digital Age

In the Free State Foundation's First Response to the Committee, we proposed the following general framework: (1) a clean slate approach to updating the Communications Act that would (2) eliminate the existing silo regime that subjects different providers of similar services to different regulatory burdens; (3) largely eliminate the overly-broad public interest standard for regulating services; (4) limit the FCC's authority to adopt broad anticipatory *ex ante* rules and confine primarily its actions to an *ex post* process based on adjudication of individual complaints alleging specific abuses of market power and consumer harm; and (5) transfer certain functions from the FCC's jurisdiction to the FTC.[249] We urged that these framework principles form the basis for a new Digital Age Communications Act. Replacement of the legacy video services regime should be part and parcel of that reform so that video services are integrated into the same regulatory framework.

First, inclusion of video within the scope of a new Digital Age Communication Act is warranted by the transition of video from largely static, one-way, analog services to increasingly interactive, digital, IP-based services. Generally, the same technological and competitive principles relevant to other advanced

---

248   See Randolph J. May, *Charting a New Constitutional Jurisprudence for the Digital Age*, 3 Charleston L. Rev. 373 (2009); Christopher S. Yoo, *The Rise and Demise of the Technology-Specific Approach to the First Amendment*, 91 Geo. L. J. 245 (2003).

249   Free State Foundation Response to Questions in the First White Paper, "Modernizing the Communications Act," (Jan. 31, 2014), at 4-5: http://www.freestatefoundation.org/images/Response_to_Questions_in_the_First_White_Paper_013114.pdf.

telecommunications services are also relevant to video services.[250] In key respects, video may be regarded as a digital app in the broadband ecosystem. Bringing video into a single, unified framework for digital services furthers the goals of policy simplicity and harmony.

Second, transition of video to digital and IP-based platforms has rendered the silo statutory structure obsolete. With competition primarily taking place across multiple digital platforms employing various technologies, and often a mix of technologies, consumers enjoy numerous choices of content providers and services. It is precisely this kind of competition among platforms for video services that has significantly reduced, and in most areas eliminated, concerns about market power and consumer harm based on scarcity, bottlenecks, or lack of alternative outlets. Different platforms for delivering video services should no longer be subject to disparate regulations simply because they fit 1990s conceptions of how different silos should operate. Regulations should no longer be geared toward managing the technical or business model particulars of broadcast TV services, cable services, or DBS services through broad prescriptive rules. Nor should Congress merely supplement the existing silo regime with new silo categories for online video distributors, mobile broadband-enabled video, or the like. Instead, Congress should replace the legacy video regulatory regime with a technologically neutral framework applicable to video services that reflects the presence of cross-platform competition.

Third, the Communications Act's public interest standard is unfit for the digital age marketplace, and it should be largely dropped.[251] By virtue of such an indeterminate delegation of authority, the FCC continues to enjoy an extraordinary degree

---

250 We recognize that circumscribed cases involving public safety or emergencies may present exceptions to application of the general principles.

251 See Randolph J. May, *The Public Interest Standard: Is It Too Indeterminate to Be Constitutional?*, 53 Fed. Comm. L. J. 427 (2001).

of discretionary power over editorial content decisions involving broadcast TV services and MVPD services, even as techno-logical advancements have transformed spectrum into a more fungible, dynamic resource. Public interest regulation of video services relying on spectrum is no longer justifiable in light of competitive and technological developments. Continued reg-ulation of video services under this standard, and the potential reinstatement of older regulations, such as the Fairness Doctrine, that have been removed but for which agency authority still exists, risks harming consumers by selectively burdening some media outlets and by discouraging the development of innovative business models.

The public interest standard is so vague that it easily can be criticized for conferring too much discretion on the agency without sufficient direction from Congress.[252] In the video con-text, the public interest standard poses acute First Amendment problems. Over the years, the FCC has invoked the public interest delegation to impose content-based regulations on broadcast TV services. Even putting aside doubts whether such content regula-tion was ever justified, technological advancements and compet-itive marketplace developments provide no persuasive rationale for subjecting broadcast TV services to a greater degree of First Amendment restriction compared to other video services and media outlets. A First Amendment-compatible approach would put broadcasters, as well as other digital media purveyors, includ-ing cable, satellite, wireless, and broadband Internet providers, on par with the First Amendment protections traditionally enjoyed by the print media.[253]

Fourth, prescriptive rules for various types of video services

---

252 See Randolph J. May, *The Public Interest Standard: Is It Too Indeterminate to Be Constitutional?*, 53 Fed. Comm. L. J. 427 (2001).

253 See Randolph J. May, *Charting a New Constitutional Jurisprudence for the Digital Age*, 3 Charleston L. Rev. 373 (2009); Christopher S. Yoo, *The Rise and Demise of the Technology-Specific Approach to the First Amendment*, 91 Geo. L. J. 245 (2003).

are unsuited to competitive conditions in today's digital market for video services. A simplified, clean slate approach should replace those rules with a case-by-case adjudicatory process tied to actual findings of market power and consumer harm. In the Free State Foundation's First Response to the Committee, we recommended Congress adopt "a competition-based standard that directs the FCC generally to undertake an antitrust-like economic analysis when it engages in regulatory activity that is subject to its jurisdiction."[254] We elaborated on the market power standard and accompanying process for case-by-case adjudication in the Free State Foundation's Third Response to the Committee:

> Regulatory prohibitions and sanctions under the new Communications Act should generally be accomplished through focused adjudicatory proceedings. The filing of individual complaints, whether by consumers or market rivals, should contain specific allegations of abuse of market power. The burden should rest on complainants to demonstrate the need for regulatory intervention by clear and convincing evidence of anticompetitive conduct and its likely resulting harm. Any regulatory intervention by the FCC should thus normally be tied to a finding of a threat of market power abuse and a concomitant threat of consumer harm. Furthermore, due to the dynamism that characterizes the modern communications marketplace, these allegations of market failure should show more than some transitory failure that can be met by targeted responses of other market participants. Therefore, any allegations of market failure

---

254 FSF's First Response, at 12.

should be "non-transitory" in order to trigger a Commission response.[255]

This same market power framework and case-by-case approach should apply to video services, replacing the silo approach and public interest standard governing video services today. In other words, the existing legacy video regulatory apparatus that presumes regulatory intervention is the norm and its sector-specific or technology-specific rules regarding video content delivery should be eliminated. A new framework should be established that is applicable to all video services in the digital marketplace, the organizing principle of which is a rebuttable presumption that runs in favor of marketplace freedom and against regulatory intervention in the video market.

In establishing a market power standard for video services, one modest agency precedent for Congress to consider is the FCC's *Program Access Order* (2012).[256] In that order, the FCC replaced its ban on exclusive contracts by vertically-integrated cable programmers with a rebuttable presumption of market competitiveness, albeit with extra qualifications attached. A market power analytical framework could build in part on the approach adopted in the *Program Access Order* by employing a more straightforward deregulatory presumption to apply to all video services.

Congress can also draw upon useful insight offered in a judicial context. Section 616 of the 1992 Cable Act contains program carriage requirements restricting MVPD conduct that will "unreasonably restrain the ability of an unaffiliated video programming vendor to compete fairly by discriminating in video programming

---

255   Free State Foundation Response to Questions in the Third White Paper, "Competition Policy and the Role of the Federal Communications Commission" (June 13, 2014), at http://www.freestatefoundation.org/images/Response_to_Questions_in_the_Third_White_Paper_061314.pdf.

256   *In re* Revision of the Commission's Program Access Rules, *Report and Order and Further Notice of Proposed Rulemaking* ("*Program Access Order*"), 27 FCC Rcd. 12619-37 (2010).

distribution on the basis of affiliation or non-affiliation of vendors in the selection, terms, or conditions for carriage." In *Comcast v. FCC* (2013), involving a program carriage complaint filed by the Tennis Channel, Judge Brett Kavanaugh interpreted Section 616 in light of antitrust law's understanding of the term "unreasonably restrain":

> Because Section 616 incorporates antitrust principles and because antitrust law holds that vertical integration and vertical contracts are potentially problematic only when a firm has market power in the relevant market, it follows that Section 616 applies only when a video programming distributor has market power in the relevant market. Section 616 thus does not bar vertical integration or vertical contracts that favor affiliated video programming networks, absent a showing that the video programming distributor at least has market power in the relevant market.[257]

Significantly, Judge Kavanaugh also addressed the First Amendment implication of the government overruling video content and channel lineup decisions made by video programming distributors. He concluded that government interference with the editorial discretion of video programming distributors is only permissible where such distributors possess market power in the relevant market. Judge Kavanaugh's antitrust-based reading of Section 616 was thereby bolstered by the constitutional avoidance canon, whereby a statute susceptible to more than one reasonable construction is interpreted to avoid raising constitutional problems. In that case, First Amendment protections for editorial decisionmaking related to video programming

---

257   717 F.3d 982, 991 (D.C. Cir. 2013) (Kavanaugh, J., concurring).

tipped the scales in favor of free speech in the absence of a market power.

Fifth, jurisdiction over consumer privacy matters related to digital video services should be turned over to the Federal Trade Commission (FTC). As part of that reform, existing FCC authority over cable subscriber privacy (Section 551) and over DBS subscriber privacy (Section 338 of the Satellite Home Viewing Improvement Act) should be transferred to the FTC.

In the Free State Foundation's First Response to the Committee, we stated:

> [W]ith regard to any regulatory oversight relative to the protection of privacy or data security, even though the FCC presently has some jurisdiction in these areas, for the most part, it would be preferable to consolidate such jurisdiction in the FTC. The types of consumer protection issues most likely to arise with regard to privacy and data security are at the core of the FTC's institutional expertise. If jurisdiction over these type of matters – matters outside of the purview of traditional economic regulation of service providers – is transferred to the FTC, it is much less likely that telecom and cable services providers, on the one hand, and, say, Facebook or Twitter, on the other, will end up subject to disparate regulations in these areas.[258]

Simple and consistent rules concerning privacy of personal data are the most consumer-friendly and what consumers in converging digital markets increasingly expect. There is no basis to presume consumers want different sets of basic data privacy protections that depend upon whether they are doing business

---

258   FSF First White Paper Response, at 17.

with, say, a cable provider, a DBS provider, or an OVD. Nor is there any basis in thinking consumers want different sets of data privacy protections from a single provider of digital services, depending on whether video, voice, or other data applications are involved. Consolidating consumer privacy for video services along with other digital services within the FTC's jurisdiction would establish a common enforcer and common set of rules for data privacy policy, providing the consistency to consumers of digital services.

V. Transitional Measures for Implementing a New Framework for Video Services

Successful implementation of a market-based policy approach to video that relies on antitrust-like analysis in case-by-case adjudication will undoubtedly require various periods of transition from the current disparate regulatory requirements in order to protect established reliance interests. To ensure the eventual end to sector-specific or provider-specific regulations based on outdated snapshots of the video market and the Communications Act's antiquated silo regime, Congress should require the Commission to adopt a series of specific sunset dates by which time legacy video mandates are to be eliminated.

Sunset date announcements will crystallize the expectations of video service providers as well as video content owners, giving them a reasonable transitional period to adjust technical and business operations to a more market-oriented approach and to take into account the adjustment of existing arrangements.

Where the Communications Act grants the FCC rulemaking authority over video services that has long since gone dormant or where the FCC has already repealed relevant regulations, no transitional provisions should apply. For instance, elimination of FCC authority to reinstitute cable leased access regulations or broadcasting rules such as network non-duplication rules, syndication exclusivity rules, or sports blackout rules should be immediate. To the extent that Congress adopts the approach suggested here, which includes jettisoning basic tier cable channel requirements, basic tier cable rate regulations, program carriage mandates, and must-carry and retransmission consent rules, it will also need to consider changes in the law in related areas, such as copyright law, where the compulsory license should be reconsidered.

VI. Conclusion

As the Committee moves forward with its review and update process, including the evaluation of competition policy, we urge it to carefully consider and implement the views expressed in this Response, as well as the previous Free State Foundation Responses. We look forward to continuing to play a constructive role in this process leading to a much-needed update of the Communications Act.

CONCLUSION:
A #COMMACTUPDATE CHECKLIST

This monograph has presented a free market and rule of law-oriented communications policy framework for the Digital Age. The basic principles and policies of the clean slate approach recommended in this book are summarized as follows:

- Eliminate the "silo" regime for regulating comparable digital communications services in a disparate manner. Eliminate also the authority for the FCC to regulate "in the public interest," except for specifically identified areas such as public safety. Replace the silo regime and public interest standard with a competition standard that is generally applicable to all entities providing digital communications services subject to the FCC's jurisdiction. This new standard should rely heavily on antitrust-like microeconomic analysis and take into account increasing cross-platform, facilities-based intermodal competition, enabled by the rise of digital and Internet Protocol-based services. Under this standard, regulatory intervention is only warranted where there is convincing evidence of a market power abuse that is likely to harm consumers.

- Circumscribe substantially the FCC's authority to adopt broad anticipatory rules and instead require the FCC to rely more heavily on adjudicating individual complaints alleging specific abuses of market power and consumer harm. Require that agency rules will sunset by operation of law after a fixed number of years absent a strong showing at the sunset date that they should be continued.

- Transfer the FCC's limited jurisdiction over cable, direct broadcast satellite, and telecommunications privacy to the FTC. Consolidating consumer privacy regulation for video, voice, and other digital services within the FTC's jurisdiction will allow for a consistent set of rules for data privacy, enforced by a single agency.

- Place clear limits on the jurisdiction of the states to engage in economic regulation of communications service providers.

- Modernize U.S. spectrum policy by replacing the command-and-control framework with a market-oriented system. Initial assignments of spectrum rights should be made through unencumbered auctions, with spectrum rights freely tradable thereafter in the secondary market. The FCC's role in spectrum management should be limited to serving as a spectrum rights-clearinghouse, performing modest "zoning" functions to solve coordination problems, and addressing special needs such as public safety.

- Ensure a minimalist role for the FCC regarding IP-based network interconnection. Rather than adopt a general duty to interconnect, the law should presume that interconnection agreements between networks will be negotiated on a voluntary basis. The Commission should intervene only upon a finding that denial of interconnection poses likely harm to consumers and that market competition is inadequate to correct the problem. In rare instances when intervention is necessary, the Commission should use some form of alternative dispute resolution, such as mediation or arbitration, rather than rate case-like adjudication.

- Refocus the Universal Service Fund on its core mission of assisting consumers who cannot afford broadband access through market-based and consumer-empowering initiatives. A reformed USF program should support build-outs only in unserved areas and only through economically efficient mechanisms – such as reverse auctions – to better ensure targeted subsidy support at the lowest cost. Any program – such as Lifeline – that is maintained for those who cannot afford service must incorporate measures to prevent fraud and waste. The antiquated contribution mechanism that places a USF surcharge or effective "tax" on all interstate and international calls must be eliminated. Instead, USF subsidies should be funded through appropriations from the general treasury.

- Establish a new unified framework for all digital video services. Regulatory oversight of video services should occur through case-by-case adjudications under a competition-based standard tied to findings of market power and consumer harm. This would replace the public interest standard, except for limited circumstances involving public safety, emergency notifications, or the like. Most of the distinct prescriptive rules for video service – such as basic tier cable channel and equipment regulations, program carriage mandates, and must-carry and retransmission consent rules – should be eliminated. In eliminating old rules, reasonable transition periods should be allowed for adjusting existing arrangements and protecting reliance interests.

- Require the FCC to adhere to certain process reforms. The FCC should be required to make a reasoned determination that the potential benefits outweigh the costs before adopting any new or revised rule that poses an economically significant impact. When reviewing mergers or other major transactions, the FCC should be authorized to impose a regulatory condition on its approval only if the condition is narrowly tailored to likely harms uniquely presented by the specific transaction. The FCC should be required to publicly post the text of agenda items several days in advance of scheduled votes at Open Meetings. Also, the FCC's required annual reports on video competition, wireless competition, broadband progress, and the like should be consolidated into a single report to be transmitted to Congress and made publicly available on a biennial basis.

About the Authors

Randolph J. May
President, The Free State Foundation

Randolph May is founder and President of The Free State Foundation. Before entering the think tank world, Mr. May practiced communications, administrative, and regulatory law as a partner at major national law firms. He previously served as Associate General Counsel at the FCC. Mr. May has held numerous leadership positions in bar associations, including serving as Chair of the American Bar Association's Section of Administrative Law and Regulatory Practice and a member of the ABA's House of Delegates. He is a Senior Fellow of the Administrative Conference of the United States, and a Fellow at the National Academy of Public Administration. He has published more than one hundred and eighty articles and essays on communications, administrative, and constitutional law topics. Most recently, Mr. May is the co-author, with FSF Senior Fellow Seth Cooper, of *The Constitutional Foundations of Intellectual Property*. He is the author of *A Call for a Radical New Communications Policy: Proposals for Free Market Reform*, published in 2011, and is editor of two books, *Communications Law and Policy in the Digital Age: The Next Five Years*, published in 2012, and *New Directions in Communications Policy*, published in 2009. In addition, he is the co-editor of two other books, *Net Neutrality or Net Neutering: Should Broadband Internet Services Be Regulated?* and *Communications Deregulation and FCC Reform*. Mr. May received his B.A. from Duke University and his law degree from Duke Law School.

Seth L. Cooper
Senior Fellow, The Free State Foundation

Seth Cooper is a Senior Fellow at The Free State Foundation. His work on federal communications and technology policy at the Free State Foundation began in 2009. He previously served as Director to the Telecommunications and Information Technology Task Force at the American Legislative Exchange Council (ALEC). Mr. Cooper served as judicial clerk to the Honorable James Johnson at the Washington State Supreme Court. He has worked in law and policy staff positions at the Washington State Senate and at the Discovery Institute. He is a 2009 Lincoln Fellow at the Claremont Institute. Mr. Cooper is the co-author, with FSF President Randolph May, of *The Constitutional Foundations of Intellectual Property*. Mr. Cooper previously contributed to chapters in *Communications Law and Policy in the Digital Age* (2012). His work has also appeared in such publications as *CommLaw Conspectus*, the *Gonzaga Law Review*, the *San Jose Mercury News*, *Forbes.com*, the *Des Moines Register*, the *Baltimore Sun*, the *Washington Examiner*, the *Washington Times*, and *The Hill*. Mr. Cooper earned his B.A. from Pacific Lutheran University and received his J.D. from Seattle University School of Law.

* * *

Members of the Free State Foundation's Board of Academic Advisors Who Contributed to One or More of the Responses to the House Energy and Commerce Committee's "White Papers"

Michelle P. Connolly
Professor of the Practice, Economics Department, Duke University

Michelle Connolly is Professor of the Practice in the Economics Department at Duke University. She was the Economics Director of Duke in New York: Financial Markets and Institutions Program for 2007-2009 and the Director of EcoTeach for several years. She currently serves as the Director of the Honors Program in Economics. Professor Connolly's research and teaching focus specifically on international trade, telecommunications policy, media policy, education, growth, and development. Professor Connolly previously served as Chief Economist of the Federal Communications Commission in 2006-2007 and 2008-2009, and as an Economist for the International Research Function for the Federal Reserve Bank of New York from 1996 to 1997. Professor Connolly has testified before Congress and participated in a White House panel on spectrum issues. She graduated from Yale University and went on to earn her M.A., M.Phil., and Ph.D. in economics also from Yale University.

Richard A. Epstein
Laurence A. Tisch Professor of Law, New York University School of Law

Richard Epstein is the inaugural Laurence A. Tisch Professor of Law at New York University School of Law, and he has served as the Peter and Kirstin Bedford Senior Fellow at the Hoover Institution since 2000. Professor Epstein is also the James Parker Hall Distinguished Service Professor of Law Emeritus and a senior lecturer at the University of Chicago. From 2001 to 2010 he was a director of the John M. Olin Program in Law and Economics at the University of Chicago. Professor Epstein has been a member of the American Academy of Arts and Sciences since 1985. Professor Epstein is the author of numerous books and scholarly articles covering a wide range of legal and interdisciplinary subjects. Courses he has taught include administrative law, antitrust law, communications, constitutional law, contracts, criminal law, labor law, property, jurisprudence, and patents. He earned his B.A. from Columbia College, a B.A. (Juris.) from Oxford University, an LL.B. from Yale Law School, and he received an LL.D., h.c. from the University of Ghent, 2003.

Justin (Gus) Hurwitz
Assistant Professor of Law at the University of Nebraska College of Law and Co-Director of the Space, Cyber, and Telecom Law Program

Gus Hurwitz is an Assistant Professor of Law at the University of Nebraska College of Law. His work builds on his background in law, technology, and economics to consider the interface between law and technology and the role of regulation in high-tech industries. He has a particular expertise in telecommunications law and technology, including data- and cybersecurity. Professor Hurwitz previously was the inaugural Research Fellow at the University of Pennsylvania Law School's Center for Technology, Innovation and Competition (CTIC). From 2007-2010 he was a Trial Attorney with the U.S. Department of Justice Antitrust Division in the Telecommunications and Media Enforcement Section. Professor Hurwitz received his J.D. from the University of Chicago Law School, an M.A. in Economics from George Mason University, and a B.A. from St. John's College.

Daniel Lyons
Associate Professor, Boston College Law School

Daniel Lyons is an Associate Professor at Boston College Law School. He specializes in the areas of property, telecommunications, and administrative law. Before joining the faculty, Professor Lyons practiced energy, telecommunications, and administrative law at the firm of Munger, Tolles and Olson in Los Angeles. He also clerked for Judge Cynthia Holcomb Hall of the Ninth Circuit Court of Appeals. Professor Lyons has participated in rulemaking proceedings before both the FCC and the California Public Utilities Commission and has represented clients in federal and state litigation involving numerous regulatory issues. He has also spoken at workshops nationwide on the effects of technology convergence on telecommunications regulation. Professor Lyons received his A.B. from Harvard College and J.D. from Harvard Law School.

Bruce M. Owen
Morris M. Doyle Centennial Professor in Public Policy, Emeritus, Stanford University and Senior Fellow, SIEPR, Emeritus

Bruce Owen is the Gordon Cain Senior Fellow, Emeritus, at SIEPR. He is the Morris M. Doyle Professor in Public Policy, Emeritus, at Stanford University and former Director of the Public Policy Program and also a professor, by courtesy, of economics. Professor Owen's research is in the areas of competition and regulation policy, law and economics, telecommunications and mass media. He served as the chief economist of the Antitrust Division of the U.S. Department of Justice and, earlier, of the White House Office of Telecommunications Policy. He was a co-founder and CEO of Economists Incorporated, a Washington DC, economic consulting firm, from 1980 to 2003. Professor Owen is the author or co-author of numerous articles and eight books. Professor Owen received his B.A. from Williams College and his Ph.D. from Stanford University.

Richard J. Pierce, Jr.
Lyle T. Alverson Professor of Law, George Washington University School of Law

Richard Pierce is the Lyle T. Alverson Professor of Law at the George Washington University Law School. Professor Pierce is the most frequently cited scholar in the country in the field of administrative law and government regulation. He is a member of the Administrative Conference of the United States, and is the author or co-author of *Administrative Law Treatise* (5th ed. 2010) and *Administrative Law & Process* (5th ed. 2009), as well as numerous other books and more than 130 articles on government regulation, regulatory economics, and the effects of various forms of government intervention on the performance of markets. His books and articles have been cited in hundreds of judicial opinions, including over a dozen opinions of the U.S. Supreme Court. Professor Pierce earned his B.S. from Lehigh University and his J.D. from the University of Virginia.

Glen O. Robinson
David and Mary Harrison Distinguished Professor of Law Emeritus, University of Virginia Law School

Glen Robinson is Professor of Law Emeritus at the University of Virginia. He served as a Commissioner of the Federal Communications Commission from 1974-1976. Professor Robinson has served as a consultant to the U.S. State Department on communications matters, and in 1979 was Ambassador and U.S. Representative to the World Administrative Radio Conference in Geneva. Professor Robinson has taught and written extensively in a number of fields, including administrative law, antitrust, communications, and cyberlaw. He practiced law with the Washington, DC, firm of Covington & Burling from 1961 to 1967. Professor Robinson earned his A.B at Harvard University and his J.D. at Stanford Law School.

James B. Speta
Professor of Law and Associate Dean, Northwestern University Law School

James Speta is Class of 1940 Research Professor of Law at the Northwestern University School of Law in Chicago, where he has been on the faculty since 1999. Professor Speta teaches antitrust, intellectual property, administrative law, business associations, torts, and telecommunications and Internet policy, and has won numerous teaching awards. In addition to teaching in the Law School, Professor Speta has taught classes at the Kellogg Business School and the McCormick School of Engineering, and he is faculty director of Northwestern's Executive LLM program in Seoul, South Korea. He is of counsel at the Chicago firm of Eimer Stahl

LLP. He has argued several cases in the U.S. Courts of Appeals and the Illinois Appellate Court and has filed merits and amicus briefs in the U.S. Supreme Court, U.S. Courts of Appeals, and several state appellate courts. Prior to joining the Northwestern faculty, he practiced for five years in the Chicago office of Sidley Austin LLP, focusing on telecommunications, antitrust, and appellate litigation. Professor Speta earned his B.A. from the University of Michigan and his J.D. from the University of Michigan.

Christopher S. Yoo
John H. Chestnut Professor of Law, University of Pennsylvania Law School

Christopher Yoo is the John H. Chestnut Professor of Law, and Professor of Computer & Information Science, at the University of Pennsylvania Law School. He is the Founding Director of the Law School's Center for Technology, Innovation, and Competition. Professor Yoo has emerged as one of the nation's leading authorities on law and technology. His research focuses on how the principles of network engineering and the economics of imperfect competition can provide insights into the regulation of electronic communications. He has been a leading voice in the "network neutrality" debate that has dominated Internet policy over the past several years. He is also pursuing research on copyright theory as well as the history of presidential power. Professor Yoo has written prolifically and testified frequently before Congress, the FCC, and the FTC. He clerked for Justice Anthony Kennedy of the Supreme Court of the United States, and the Honorable A. Raymond Randolph of the United States Court of Appeals for the D.C. Circuit, Washington, DC. Professor Yoo earned his A.B. from Harvard, M.B.A. from UCLA, and J.D. from Northwestern University School of Law.

About The Free State Foundation

The Free State Foundation is a non-profit, nonpartisan think tank. Its purpose is to promote, through research and educational activities, understanding of free market, limited government, and rule of law principles at the federal level and in Maryland.

The Free State Foundation focuses on eliminating unnecessary and counterproductive regulatory mandates, especially those applicable to the communications and other high-tech industries, and on reducing overly burdensome taxes, protecting individual and economic liberty, including property rights, and making government more effective, efficient, and accountable.

Led by Randolph J. May, FSF's President, the scholars of the Free State Foundation have decades of experience in the public policy arena and academic settings promoting free markets, secure property rights, and individual liberty. Mr. May is a nationally known legal and policy expert in the field of regulation, as well as a widely recognized expert in communications, administrative, and constitutional law. FSF's scholars combine solid academic expertise with professional and practical experience in a way that makes their research especially impactful in influencing policy.

Appendix

Testimony of Randolph J. May
President, The Free State Foundation

Hearing on "Improving FCC Process"

before the

Subcommittee on Communications and Technology

Committee on Energy and Commerce

U.S. House of Representatives

July 11, 2013

Summary of the Testimony of Randolph J. May
President, The Free State Foundation

I commend the Committee for undertaking this effort to reform the FCC's processes and its decision-making approaches, and I support the proposed reforms in the Discussion Drafts. Given the increasing competiveness in the communications marketplace, FCC reforms, such as those embodied in the draft bills, are needed now more than ever.

The FCC still operates today with a pro-regulatory bent pretty much as it did in 1999 when FCC Chairman William Kennard called for the reorientation of the agency's mission to account for the increasingly competitive environment evident even then. The reforms in the draft bills, along with a few additional proposals I will suggest, would make the FCC less likely to default so often to regulatory measures, even absent clear and convincing evidence of market failure or consumer harm. In today's marketplace environment, the default position should not be regulation.

I wish to highlight here the proposed reform of the rulemaking requirements and the transaction review process because they are especially consequential. New Section 13(a)(2)(C)(iii)'s requirement that the Commission, before adopting a new or revised rule, provide a reasoned explanation why market forces and technology changes will not, within a reasonable time period, resolve the agency's concerns is particularly welcome. I urge the

Committee to go a step further to make it more difficult for the Commission to avoid the import of this provision while carrying on "business as usual." I suggest revising the provision to read: "(iii) a reasoned determination, <u>based on clear and convincing evidence</u>, that market forces or changes in technology...." This change will not prevent the Commission from adopting any new regulations, and it is not intended to do so. But, without altering the substantive criteria the bill specifies, the suggested change simply requires the agency to meet a higher evidentiary burden before adopting or revising regulations.

The provisions contained in new Section 13(k), especially the addition that would allow the Commission to condition approval of a proposed transaction only if the condition addresses a likely harm uniquely presented by the specific transaction, would go a long way toward combatting abuse of the transaction review process. Over time, the agency increasingly has abused the merger review process by delaying approval of transactions until the applicants "voluntarily" agree – usually at the "midnight hour" – to conditions not narrowly tailored to remedy a harm arising from the transaction or unique to it.

I also suggest the Committee reform the forbearance and periodic regulatory review process by, in effect, requiring a higher evidentiary burden to maintain existing regulations on the books. Absent clear and convincing evidence that the regulations at issue should be retained under the existing substantive statutory criteria, regulatory relief should be granted. Similarly, I propose adoption of a "sunset" requirement so that all rules will automatically expire after five [or X] years absent a showing, based on clear and convincing evidence, that it is necessary for such rule to remain in effect to accomplish its original objective.

Testimony of Randolph J. May
President, The Free State Foundation

Mr. Chairman and Members of the Committee, thank you for inviting me to testify. I am President of The Free State Foundation, a non-profit, nonpartisan research and educational foundation located in Rockville, Maryland. The Free State Foundation is a free market-oriented think tank that, among other things, focuses its research in the communications law and policy and administrative law and regulatory practice areas. I have been involved for thirty-five years in communications law and policy in various capacities, including having served as Associate General Counsel at the Federal Communications Commission. While I am not speaking on behalf of these organizations, by way of background I wish to note that I am a past Section Chair of the American Bar Association's Section of Administrative Law and Regulatory Practice and its representative in the ABA House of Delegates. I am currently a Public Member of the Administrative Conference of the United States and a Fellow at the National Academy of Public Administration. So, today's hearing on FCC process reform is at the core of my longstanding experience and expertise in communications law and policy and administrative law and regulatory practice.

I appreciated the opportunity to testify before this Committee a bit more than two years ago on June 22, 2011, at the hearing on "Reforming the FCC Process," and I appreciate the opportunity to testify today.

Though H. R. 3309 and H. R. 3310 both passed the House, unfortunately they died in the Senate. I want to begin by saying that reform measures like those embodied in those bills and the present Discussion Drafts, or very similar ones, are needed now more than ever. In my June 2011 testimony, I generally supported the proposed reforms, and I do so again today. I do so because the Federal Communications Commission needs to change in a way so that, in today's generally dynamic, competitive communications marketplace environment, it will be less prone to continue on its course of too often defaulting to regulatory solutions, even when there is no clear and convincing evidence of market failure or consumer harm.

In addition to supporting the Discussion Drafts, including the few changes that are included in the draft bills that were not part of H. R. 3309 and 3310, I want to suggest a few additional reform proposals for consideration as well. These proposals, though requiring only relatively small revisions to the language of the Communications Act, would be useful as complements to the measures proposed in the Discussion Drafts as a means of requiring the FCC to eliminate or reduce unnecessary regulation. And this point is key: They do so not by altering the substantive regulatory criteria presently in the Communications Act relating to protecting consumers and the public interest, but rather by establishing higher evidentiary burdens the Commission would be required to meet in deciding whether to maintain existing regulations or adopt new ones.

At the outset of my testimony two years ago, to set the stage for explaining why Congress should adopt FCC reform measures, I presented statements made over a decade ago by two different FCC commissioners. In August 1999, FCC Chairman William Kennard released a strategic plan entitled, "A New FCC for the 21st Century." The plan's first four sentences read:

"In five years, we expect U.S communications markets to be characterized predominately by vigorous competition that will greatly reduce the need for direct regulation. The advent of In-ternet-based and other new technology-driven communications services will continue to erode the traditional regulatory distinc-tions between different sectors of the communications industry. As a result, over the next five years, the FCC must wisely manage the transition from an industry regulator to a market facilitator. The FCC as we know it today will be very different in both struc-ture and mission."

In December 2000, then-FCC Commissioner (soon-to-be FCC Chairman) Michael Powell, in his "Great Digital Broadband Migration" speech, said: "Our bureaucratic process is too slow to respond to the challenges of Internet time. One way to do so is to clear away the regulatory underbrush to bring greater certainty and regulatory simplicity to the market."

These statements by two FCC Chairman, one a Democrat and the other a Republican, still provide a most useful frame for thinking about today's topic. Without belaboring the point now with all the latest marketplace facts and figures, we should be able to agree, regardless of party identification, that, as Bill Kennard predicted they would be, U.S. communications markets are now "characterized predominately by vigorous competition."

Despite the fact that the communications marketplace incontrovertibly is characterized by much more dynamism and competition now than at the turn of the century – and that econ-omists and regulatory experts agree that increased marketplace competition generally should supplant the need for regulation – the FCC's staffing levels have maintained essentially level since 2000, and the amount the agency spends on regulation has in-creased substantially during that period. In both 2000 and today, the FCC's FTE employee count stands roughly in the 1900 range. And from 2000 to 2012, based on data extracted from the *Budget of the United States Government* and compiled by the Weiden-baum Center on the Economy, Government and Public Policy at Washington University and the George Washington University

Regulatory Studies Center, the amount the FCC spends on regulatory activity (in constant 2005 dollars) has increased from $303 million to $392 million.

While these figures are not intended to – and don't – show the benefits and costs of any particular regulations or suggest that regulation is not still appropriate in particular market segments or areas, they do suggest that the FCC still operates today with a pro-regulatory bent pretty much as it did in 1999 when Bill Kennard called for the reorientation of the agency's mission to account for the increasingly competitive environment and in 2000 when Michael Powell urged that the agency remake itself so that it can respond to the challenges of "Internet time."

Hence the need now for Congress to adopt meaningful FCC regulatory reform measures.

The Federal Communications Commission Process Reform Act

I support the proposals in the Process Reform Act Discussion Draft and commend the Committee for undertaking this effort. In my testimony, I just want to highlight here the provisions that I think are most important, suggest three relatively minor revisions to the language of the draft, and then propose three additional measures that I believe are consistent with the FCC reform the Committee is trying to accomplish.

Section 13(a) – Rulemaking Reforms. In light of what I have already said concerning the dynamic, generally competitive state of the communications marketplace, I want highlight new Section 13(a) relating to the adoption of new or revised FCC rules and especially Section 13(a)(2)(C). Section 13(a)(2)(C)'s requirement, regarding adoption or revision of a rule that may have an economically significant impact, that the Commission must (i) identify and analyze the market failure and actual consumer harm the rule addresses; (ii) make a reasoned determination that the rule's benefits justify the costs; and (iii) make a reasoned determination that market forces and changes in technology are unlikely to resolve within a reasonable period of time the problem the Commission intends the rule to address is particularly important. As I have explained, despite the dramatic marketplace changes that have occurred over the past couple of decades, the Commission still too often defaults to regulatory solutions when they are not justified. Requiring the Commission to perform the identification and analysis and to make the determinations specified in Section 13(a)(2)(C) should be helpful in combatting the FCC's tendency to default to regulatory solutions without undertaking rigorous economic analysis, considering the cost and benefits of regulations, and evaluating marketplace conditions.

Section 13(a)(2)(C)(iii)'s requirement is a very welcome addition to the Process Reform Act that was not present in H. R. 3309. Requiring the Commission to explain in a reasoned way why market forces and technology changes will not, within a reasonable period of time, resolve the agency's concerns is consistent with recommendations I have made in the past. While the addition is positive, I would urge the Committee to go a step further in order to make it more difficult for the Commission to avoid the import of this provision while carrying on "business as usual." I suggest revising the provision to read: "(iii) a reasoned determination, based on clear and convincing evidence, that market forces or changes in technology…." This change will not prevent the Commission from adopting any new regulations, and, indeed, it is not intended to do so. Without altering the substantive criteria that the bill specifies the FCC must consider, the suggested change simply requires the agency to meet a higher evidentiary burden before adopting or revising regulations.

Section 13(c) – Sunshine Act Reforms. I endorse the proposed changes to the Sunshine Act. Currently, the Act's strictures, without any meaningful public benefit, prevent the agency's five commissioners from engaging in the type of collaborative discussions that may lead to more reasoned decision-making. And they inhibit the development of greater collegiality among the commissioners, which itself may contribute to more effective functioning of a multi-member commission. I led a study in 1995 on this subject for the Administrative Conference of the United States, the results of which are published in 49 *Administrative Law Review* 415, which made recommendations somewhat similar to the draft bill's proposals.

Section 13(k) – Transaction Review Process Reforms. As I testified in 2011, the new Section 13(k) provision that would reform the Commission's transaction review process is as important as any other in the bill in light of the abuse of the process for many years now. The agency often imposes extraneous conditions -- that is, conditions not related to any alleged harms caused by the proposed transaction – after they are "volunteered" at the last-minute by transaction applicants anxious to get their deal done. The bill's requirement that any condition imposed be narrowly tailored to remedy a transaction-specific harm, coupled with the provision that the Commission may not consider a voluntary commitment offered by a transaction applicant unless the agency could adopt a rule to the same effect, go a long way to reforming the review process. But the Discussion Draft now contains an additional provision, Section 13(k)(1)(c), that allows the Commission to condition approval of the transaction only if the condition addresses a likely harm uniquely presented by the specific transaction. This is a very good addition that will reduce the wiggle room for the Commission to continue abusing the transaction review process by imposing conditions that, if imposed at all, should be imposed only on an industry-wide basis in generic rulemaking proceedings.

I first suggested reforms exactly along these lines, including the new addition, in an essay entitled "Any Volunteers?" in the March 6, 2000 edition of *Legal Times*, so I am very pleased with the transaction review proposal. And as said in the *Legal Times* essay, and in my testimony in 2011, my own preference would be to go even further to reduce the substantial overlap in work and expenditure of resources that now occurs when the antitrust agencies and the FCC engage in a substantial duplication of effort. I would place primary responsibility for assessing the competitive impact of proposed transactions in the hands of the Department of Justice and the Federal Trade Commission, the agencies with the most expertise in this area. The FCC's primary responsibility then would be to ensure the applicants are in compliance with all rules and statutory requirements.

Other Provisions. I support the provision that would require publication of the text of agenda items in advance of an open meeting so that the public has the opportunity to review the text before a vote is taken. Before each and every item is considered by the commissioners at a public meeting the staff requests and is granted so-called "editorial privileges." Because the public does not have the text upon which the commissioners are voting, the public has no way of knowing the extent to which a draft order is actually changed – that is, the extent to which editorial privileges are exercised and for what purpose – after a vote but before the item *eventually* is released as a final order. I emphasize "eventually" in the previous sentence because, as this Committee knows, there have been some lengthy delays in releasing orders to the public after they supposedly have been approved at open meetings. Thus, I support the provision that requires the Commission to publish each order or other action no later than 7 days after the date of adoption, or at least within some reasonably short period.

Along the same lines, I support the provision that requires the Commission to establish deadlines for Commission orders and other actions and to release promptly certain identified reports. And I support the provision in the draft bill that provides that the Commission may not rely in any order or decision on any statistical report, report to Congress, or *ex parte* communication unless the public has been afforded adequate notice and opportunity to comment. A large amount of material, including studies, articles, and reports, was "dumped" into the docket of the net neutrality proceeding only a few days before the Commission adopted a draft order citing many of these documents. This last-minute "data dump" made it difficult, if not impossible, for the public to review and comment on the new material in the docket.

New Section 13(e) requiring brief advance notice to the commissioners of an action proposed to be taken on delegated authority and allowing two or more commissioners to require that the action be brought before the full Commission makes sense. The Committee might wish to consider formalizing somewhat the objection procedure to avoid confusion. For example, Section 13(e) might be revised to provide that "2 or more Commissioners may <u>file an objection in writing to</u> prevent an order…."

New Section 13(l) requires the Commission to publish certain information on its website, including the total number of its full-time equivalent employees. I think this is useful information, but, as a complement, it would be useful if the Commission were required to provide information concerning the number of contractors it retains to perform work for the Commission, for what purpose, the length of the contracts, and the material terms of the contract.

Additional Reform Recommendations for the Process Reform Act

As I said early in my testimony, the reality is, as FCC Chairman William Kennard predicted in 1999, most segments of the communications marketplace are now effectively competitive and have been so for a number of years. Indeed, when Congress passed the landmark Telecommunication Act of 1996, it anticipated the development of a competitive marketplace that would lead to less regulation. In the statute's preamble, Congress stated that it intended for the FCC to "promote competition and reduce regulation." And in the principal legislative report accompanying the 1996 Act, Congress stated its intent to provide for a "de-regulatory national policy framework." In other words, Congress understood that the development of more competition and more consumer choice should lead to reduced regulation.

But the fact is that the FCC has not done nearly enough in the 17 years since the 1996 Act's adoption to "reduce regulation" and provide a "de-regulatory" framework. Whatever the reason, the key point is that a fix is needed. As I have said, the Discussion Drafts are very commendable. But, in my view, there are a few additional reform measures that should be included in the bills to more effectively ensure that the FCC does not maintain in force existing regulations, or adopt new regulations, that are not necessary to protect consumers from harm. Enactment of these measures would require only modest changes in the Communications Act's language, and I hope the Committee will consider including them in the bills so as to better effectuate what Congress intended to be the 1996 Act's deregulatory intent.

The Forbearance Relief and Periodic Regulatory Review Provisions

The 1996 Act introduced two related deregulatory tools rarely – if ever -- found in other significant statutes governing regulatory agencies. The first provision, Section 10 of the Communications Act, titled "Competition in Provision of Telecommunications Service," states the Commission "shall forbear" from enforcing any regulation or statutory provision if the agency determines, taking into account competitive market conditions, that such regulation or statutory provision is not necessary to ensure that telecommunications providers' charges and practices are reasonable, or necessary to protect consumers or the public interest. The second provision, Section 11 in the Act, titled "Regulatory Reform," requires periodic reviews of regulations so that the Commission may determine "whether any such regulation is no longer in the public interest as a result of meaningful economic competition between providers of such service." The agency is required to repeal or modify any regulation it determines to be no longer in the public interest.

While these two provisions obviously were added as tools to be used to reduce regulation in the face of developing competition, the FCC has utilized them too sparingly. In its forbearance and regulatory review rulings, the agency generally takes a very cramped view of evidence submitted concerning marketplace competition — for example, refusing to acknowledge that wireless operators compete with wireline companies by offering substitutable services, or that potential entrants exert market discipline on existing competitors, or that present market shares are not as meaningful in a technologically dynamic, rapidly changing marketplace as they may be in a static one.

The Section 10 forbearance and Section 11 periodic review provisions can be made more effective deregulatory tools simply by adding language that requires the FCC to presume, absent clear and convincing evidence to the contrary, that the consumer protection and public interest criteria for granting regulatory relief have been satisfied. And the two regulatory relief provisions should be made applicable to all entities subject to FCC regulation, not just telecommunications providers.

This sentence could be added at the end of Section 10(a): "In making the foregoing determinations, absent clear and convincing evidence to the contrary, the Commission shall presume that enforcement of such regulation or provision is not necessary to ensure that an entity's charges or practices are not unreasonable or unreasonably discriminatory or necessary for the protection of consumers and is consistent with the public interest." Similarly, a sentence could be added to the Section 11 regulatory review provision which states: "In making the foregoing determination, absent clear and convincing evidence to the contrary, the Commission shall presume that such regulation is no longer necessary in the public interest as a result of meaningful competition between providers of such service."

The specified consumer protection and public interest criteria would not be changed. But by establishing a rebuttable evidentiary presumption, in carrying out its duties under these two provisions, only those regulations supported by clear evidence that the substantive criteria have not been met would be retained. It is possible the FCC might seek to ignore or skew evidence in order to avoid reducing regulation, but I assume the agency's good faith in following congressional directives – and, in any event, the agency's decisions are subject to review by the courts.

Limitation on General Rulemaking Authority

Section 201(b) of the Communications Act provides that the Commission "may prescribe such rules and regulations as may be necessary in the public interest to carry out the provisions of this Act." This is the grant of rulemaking authority that was relied on so heavily by Justice Antonin Scalia in the recent *City of Arlington* v. *FCC* case as a reason for granting the agency such broad sway for so-called *Chevron* deference. When an agency receives *Chevron* deference upon judicial review, the agency's interpretation of its statutory authority is entitled to "controlling weight" and must be upheld unless it is unreasonable. A simple proviso could be added at the end of Section 201(b) to the effect that, before adopting a rule, "the Commission must determine, based on a showing of clear and convincing evidence presented in the rulemaking proceeding, that marketplace competition is not sufficient adequately to protect consumers from harm." This change would not prevent the Commission from adopting new regulations. Rather it would simply require the Commission to meet a higher evidentiary burden before doing so.

Sunset Requirement for Agency Regulations

Congress could add a general sunset provision to the Communications Act that provides that all rules will expire automatically after five [or X] years absent a showing by the Commission, based on clear and convincing evidence compiled after public notice and comment, that it is necessary for such rule to remain in effect to accomplish its original objective or objectives. Again, this sunset provision would not dictate that regulations expire. Instead, it would require that the agency bear the evidentiary burden of showing that such regulations be retained.

None of these proposals I have suggested would change the substantive regulatory criteria, such as protecting consumers and the public interest, that presently are in the Communications Act. Rather in each instance they simply require the Commission to show by clear and convincing evidence that existing regulations should remain on the books or that new regulations should be adopted. I urge the Committee to consider these proposals in conjunction with the other worthwhile reform measures it is considering.

The FCC Consolidated Reporting Act

I wholeheartedly support new Section 14, the proposed Federal Communications Commission Consolidated Reporting Act of 2013. The required consolidated report would replace the myriad of existing sector and technology-specific marketplace reports that the Commission is now required to compile on a periodic basis. Consolidation of the various competition/marketplace status reports should help reduce the agency's workload somewhat because there necessarily is some inherent duplication in producing the half dozen or more separate reports. But, more importantly, the requirement to produce a consolidated report should steer the Commission away from its pronounced tendency to view the separate technology-based services as confined to their own "smokestacks" and non-competitive with each other. In today's competitive digital services environment characterized by convergence, adhering to the "smokestack" view inherently neglects marketplace realities. For example, the Commission still refuses to acknowledge the extent to which wireless services compete with wireline services, even though nearly 40% of U.S. households have abandoned landline telephone service.

The draft bill requires the Commission to assess competition in the communications marketplace, taking into account all the various services and technologies, and it specifically directs the agency "to consider the effect of intermodal competition, facilities-based competition, and competition from new and emergent communications services, including the provision of content and communications using the Internet." This requirement is especially important as part of the necessary effort to get the FCC to take a more realistic, economically rigorous, view of the extent to which competition now prevails in the communications marketplace.

Thank you for giving me the opportunity to testify today. I will be pleased to answer any questions.

www.ingramcontent.com/pod-product-compliance
Lightning Source LLC
Chambersburg PA
CBHW060324030426
42336CB00011B/1196